eTELEMED 2018

The Tenth International Conference on eHealth, Telemedicine, and Social Medicine

March 25 – 29, 2018

Rome, Italy

eTELEMED 2018 Editors

Yoshitoshi Murata, Iwate Prefectural University, Japan
Shada Abdullah AlSalamah, Massachusetts Institute of Technology, USA / King Saud University (KSU), Saudi Arabia
Hessah A. AlSalamah, King Saud University (KSU), Saudi Arabia
Giuseppe Quero, University of Sannio, Italy
Bobby Gheorghiu, Evaluation Services, Canada Health Infoway/ Inforoute Santé du Canada, Canada
Hans C. Ossebaard, National Health Care Institute - Diemen, The Netherlands

Printed from e-media with permission by:

Curran Associates, Inc.
57 Morehouse Lane
Red Hook, NY 12571

Some format issues inherent in the e-media version may also appear in this print version.

Printed by Curran Associates, Inc. (2018)

International Academy, Research, and Industry Association (IARIA)
412 Derby Way
Wilmington, DE 19810

Phone: (408) 893-6407
Fax: (408) 527-6351

petre@iaria.org

Additional copies of this publication are available from:

Curran Associates, Inc.
57 Morehouse Lane
Red Hook, NY 12571 USA
Phone: 845-758-0400
Fax: 845-758-2633
Email: curran@proceedings.com
Web: www.proceedings.com

eTELEMED 2018

Forward

The tenth edition of The International Conference on eHealth, Telemedicine, and Social Medicine (eTELEMED 2018), held in Rome, Italy, March 25 - 29, 2018, considered advances in techniques, services, and applications dedicated to a global approach of eHealth.

Development of wireless homecare, of special types of communications with patient data, of videoconferencing and telepresence, and the progress in image processing and date protection increased the eHealth applications and services, and extended Internet-based patient coverage areas. Social and economic aspects as well as the integration of classical systems with the telemedicine systems are still challenging issues.

eTELEMED 2018 provided a forum where researchers were able to present recent research results and new research problems and directions related to them. The topics covered aspects from classical medicine and eHealth integration, systems and communication, devices, and applications.

We take this opportunity to thank all the members of the eTELEMED 2018 Technical Program Committee as well as the numerous reviewers. The creation of such a broad and high-quality conference program would not have been possible without their involvement. We also kindly thank all the authors who dedicated much of their time and efforts to contribute to the eTELEMED 2018. We truly believe that, thanks to all these efforts, the final conference program consists of top quality contributions.

This event could also not have been a reality without the support of many individuals, organizations, and sponsors. We are grateful to the members of the eTELEMED 2018 organizing committee for their help in handling the logistics and for their work to make this professional meeting a success.

We hope that eTELEMED 2018 was a successful international forum for the exchange of ideas and results between academia and industry and for the promotion of progress in eHealth and Telemedicine research.

We also hope that Rome provided a pleasant environment during the conference and everyone saved some time for exploring this beautiful city.

eTELEMED 2018 Chairs

eTELEMED Steering Committee
Lisette Van Gemert-Pijnen, University of Twente - Enschede, the Netherlands
Marike Hettinga, Windesheim University of Applied Sciences, the Netherlands
Joseph Finkelstein, Center for Bioinformatics and Health Analytics - Columbia University, USA
Åsa Smedberg, Stockholm University, Sweden
Yoshitoshi Murata, Iwate Prefectural University, Japan
Anthony Glascock, Drexel University, USA
Britt Östlund, Royal Institute of Technology (KTH), Sweden

eTELEMED Industry/Research Advisory Committee
Anne G. Ekeland, Norwegian Centre for e-Health Research, Norway
Salah H. Mandil, eStrategies & eHealth for WHO and ITU - Geneva, Switzerland
Bobby Gheorghiu, Canada Health Infoway/Inforoute Santé du Canada, Canada

eTELEMED 2018

COMMITTEE

eTELEMED Steering Committee

Lisette Van Gemert-Pijnen, University of Twente - Enschede, the Netherlands
Marike Hettinga, Windesheim University of Applied Sciences, the Netherlands
Joseph Finkelstein, Center for Bioinformatics and Health Analytics - Columbia University, USA
Åsa Smedberg, Stockholm University, Sweden
Yoshitoshi Murata, Iwate Prefectural University, Japan
Anthony Glascock, Drexel University, USA
Britt Östlund, Royal Institute of Technology (KTH), Sweden

eTELEMED Industry/Research Advisory Committee

Anne G. Ekeland, Norwegian Centre for e-Health Research, Norway
Salah H. Mandil, eStrategies & eHealth for WHO and ITU - Geneva, Switzerland
Bobby Gheorghiu, Canada Health Infoway/Inforoute Santé du Canada, Canada

eTELEMED 2018 Technical Program Committee

Nitin Agarwal, University of Arkansas at Little Rock, USA
Davide Aloini, University of Pisa, Italy
Hessah A. Alsalamah, King Saud University, Saudi Arabia
James G. Anderson, Purdue University, USA
Adina Astilean, Technical University of Cluj-Napoca, Romania
Luigi Benedicenti, University of Regina, Canada
Javad Birjandtalab, The University of Texas at Dallas, USA
Andrzej Cacko, Medical University of Warsaw, Poland
Manuel Campos Martínez, University of Murcia, Spain
Elena Cardillo, Institute of Informatics and Telematics - National Research Council, Italy
Maria Teresa Chiaravalloti, National Research Council | Institute of Informatics and Telematics, Rende, Italy
Marcos C. d'Ornellas, Universidade Federal de Santa Maria, Brazil
Cecilia Dias Flores, Universidade Federal de Ciências da Saúde de Porto Alegre, Brazil
Talko Dijkhuis, Hanze University of Applied Sciences, Netherlands
Aranka Do, Hanze University of Applied Sciences, Netherlands
Dmitriy Dubovitskiy, De Montfort University, UK
Georg Duftschmid, Medical University Vienna, Austria
Leila Eadie, University of Aberdeen, UK
Claudio Eccher, FBK-Fondazione Bruno Kessler, Italy
Øystein Eiring, National Institute of Public Health, Norway
Anne G. Ekeland, Norwegian Centre for e-Health Research, Norway
Christo El Morr, York University, Canada
Henrik Eriksson, Linköping University, Sweden
Ivan Evgeniev, Technical University of Sofia, Bulgaria
Manuel Filipe Santos, Universidade do Minho | Research Centre Algoritmi, Portugal
Joseph Finkelstein, Center for Bioinformatics and Health Analytics - Columbia University, USA

Audrey Dunn Galvin, University College Cork, UK
Niels F. Garmann-Johnsen, University of Agder, Norway
Anthony Gelibert, LCIS, France
Gersende Georg, Haute Autorité de Santé, France
Bobby Gheorghiu, Canada Health Infoway, Canada
Piero Giacomelli, SPAC SPA - Arzignano (Vicenza), Italy
Anthony Glascock, Drexel University, USA
Wojciech Glinkowski, Polish Telemedicine Society / Center of Excellence "TeleOrto", Poland
Jose A. Gonzalez, University of Sheffield, UK
Adela Grando, Arizona State University, USA
Conceição Granja, Norwegian Centre for eHealth Research | University Hospital of North Norway, Norway
David Greenhalgh, University of Strathclyde, Scotland
Päivi Hämäläinen, National Institute for Health and Welfare, Finland
Yongqun "Oliver" He, University of Michigan - Medical School, USA
Oliver Heinze, University Hospital Heidelberg, Germany
Marike Hettinga, Windesheim University of Applied Sciences, Netherlands
Kendall Ho, University of British Columbia / Vancouver General Hospital, Canada
Yan Hu, Blekinge Institute of Technology, Sweden
Chun-Hsi Huang, University of Connecticut, USA
Ashutosh Jadhav, IBM Research - Almaden, USA
Eleni Kaldoudi, EAMBES / Democritus University of Thrace, Greece
Bridget Kane, Karlstad University Business School, Sweden
Saif Khairat, University of North Carolina at Chapel Hill, USA
Ali Khan, Bremen University, Germany
Pinar Kirci, Istanbul University, Turkey
Spyros Kitsiou, University of Illinois at Chicago, USA
Haridimos Kondylakis, Institute of Computer Science | Foundation for Research and Technology Hellas, Greece
Vinay Kumar, Thapar University, Patiala, India
Lenka Lhotskam, Czech Institute of Informatics, Robotics and Cybernetics - Czech Technical University in Prague, Czech Republic
Sarah N. Lim Choi Keung, University of Warwick, UK
Margarita Loyola, Telehealth - Island Health, Canada
Flora Malamateniou, University of Piraeus, Greece
José Manuel Machado, University of Minho, Portugal
Salah H. Mandil, eStrategies & eHealth for WHO and ITU - Geneva, Switzerland
Catalina Martínez Costa, Institute for Medical Informatics, Statistics and Documentation - Medical University of Graz, Austria
Janet Mattsson, The Red Cross University College, Sweden
Nikola Milosevic, University of Manchester, UK
Yoshitoshi Murata, Iwate Prefectural University, Japan
Wendy Oude Nijeweme - d'Hollosy, University of Twente, Netherlands
Lambert J.M. Nieuwenhuis, University of Twente / Fontys International Business School, Netherlands
Nicholas A. I. Omoregbe, Covenant University, Nigeria
Britt Östlund, Royal Institute of Technology (KTH), Sweden
Fulvio Patara, University of Florence, Italy
Paolo Perego, Politecnico di Milano, Italy

Vitor Pinheiro de Almeida, IBM Research, Brazil
Isabel Pirillo, University of Camerino, Italy
Temiloluwa Prioleau, Rice University, Houston, USA
Giuseppe Quero, University of Sannio, Italy
Haider Raad, Xavier University, Cincinnati, USA
Humairat Rahman, University of New England, USA
Elisabeth Rakus-Andersson, Blekinge Institute of Technology, Sweden
David Riaño, Universitat Rovira i Virgili, Spain
Rainer Röhrig, Carl von Ossietzky University Oldenburg, Germany
Giovanni Saggio, University of Rome "Tor Vergata", Italy
Sanna Salanterä, University of Turku, Finland
Abdel-Badeeh M. Salem, Ain Shams University Abbasia, Cairo, Egypt
Johannes Sametinger, Johannes Kepler Universität Linz, Austria
Hamman Samuel, University of Alberta, Canada
Fabien Scalzo, University of California Los Angeles (UCLA), USA
Jacob Scharcanski, UFRGS - Universidade Federal do Rio Grande do Sul, Brazil
Artur Serrano, Norwegian Centre for eHealth Research, University Hospital of North Norway, Norway
Mohammad Shafahi, Utrecht University / University of Amsterdam, The Netherlands
Berglind Smaradottir, University of Agder, Norway
Åsa Smedberg, Stockholm University, Sweden
Sanjay Sood, Centre for Development of Advanced Computing (C-DAC), Mohali, India
Alessandro Stefanini, University of Rome Tor Vergata, Italy
Karl A. Stroetmann, empirica Communication & Technology Research, Germany / University of Victoria, Canada
Judith Tabolt Matthews, University of Pittsburgh, USA
Gianluigi Tiberi, UBT srl, Italy
Vicente Traver Salcedo, ITACA - Universitat Politècnica de València, Spain
Ceara Treacy, Dundalk Institute of Technology, Ireland
Masatsugu Tsuji, Kobe International University, Japan
Bhekisipho Twala, University of Johannesburg, South Africa
Gary Ushaw, Newcastle University, UK
Aristides Vagelatos, CTI&P, Athens, Greece
Lisette Van Gemert-Pijnen, University of Twente - Enschede, The Netherlands
George Vassilacopoulos, University of Piraeus, Greece
Miriam Vollenbroek-Hutten, University of Twente, Netherlands
Rami Yared, Bishop's University, Sherbrooke, Canada
Inas A. Yassine, Cairo University, Egypt
Sule Yildirim-Yayilgan, Norwegian University of Science and Technology, Norway
Zeljko Zilic, McGill University, Canada

Copyright Information

Table of Contents

Overall Quality of Life and General Health - Changes Related to the Retirement Threshold

Agnieszka Bielińska*, Mikołaj Majkowicz‖ **, Piotr Wąż§ and Dorota Bielińska-Wąż¶
* Department of Quality of Life Research, Medical University of Gdańsk
Gdańsk, Poland
Email: agnieszka.bielinska1984@gmail.com
‖ Department of Quality of Life Research, Medical University of Gdańsk
Gdańsk, Poland
Email: mmajk@gumed.edu.pl
** Department of Public Health, Pomeranian University in Słupsk
Słupsk, Poland
§ Department of Nuclear Medicine, Medical University of Gdańsk
Gdańsk, Poland
Email: phwaz@gumed.edu.pl
¶ Department of Radiological Informatics and Statistics, Medical University of Gdańsk
Gdańsk, Poland
Email: djwaz@gumed.edu.pl

***Abstract*—In the present work, changes of the overall quality of life and general health related to the retirement threshold are studied. In particular, the influence of different factors on the satisfaction with health and the quality of life is evaluated. The results show that education, job position and activity, as e.g., the attendance of the University of the Third Age, strongly influence the satisfaction level on the overall quality of life and general health after the retirement threshold.**

Keywords–Social medicine; Medical informatics; Correspondence analysis; University of the Third Age (U3A); World Health Organization Quality of Life-BREF (WHOQOL-BREF)

I. INTRODUCTION

Usually, critical moments in our life influence our lifestyle and health. One of the most important moments is the retirement. Changes in health related to the retirement threshold have been observed in numerous studies [1]-[6]. In particular, changes of the physical activity [2][7]-[17] and changes of the quality of life in different domains [18]-[22] have been broadly discussed. Some problems may be specific for particular countries, as for example depression related to the loss of the employment-based insurance [3]. Aging is a natural process, however different factors may influence it. In particular, a high level of education is known to have a positive influence on health and on the quality of life. Recently, Universities of the Third Age (U3A) became popular, and their positive influence is broadly discussed [23]-[28].

In this article, we focus on the problem from the Polish perspective. We concentrate on changes in the overall quality of life and general health related to the retirement threshold.

The article is organized as follows: in Section II we describe the applied methods, in Section III we describe the groups of the individuals who participated in the studies and the results related to the overall quality of life and to the general health. A summary is given in Section IV.

II. METHODS

The studies are concerned with the Polish society. In order to check changes of the quality of life and of health after the retirement threshold, we compare the results for the employees and for the retirees. The quality of life and health have been determined by using the Polish version of the World Health Organization Quality of Life-BREF (WHOQOL-BREF) questionnaire [29]. Additionally, each respondent answered several question about his/her education and the job position during the employment.

For the graphical representation of the results we apply the correspondence analysis originally introduced by Hirschfeld [30] and later improved by many authors. In this method, one obtains maps in which the objects under consideration are represented by points. In the present studies, the points represent particular groups of individuals and the answers to the questions. This kind of approach has an interdisciplinary character. In particular, we have created an analog of this kind of method in the theory of molecular similarity [31] and in bioinformatics [32].

III. RESULTS AND DISCUSSION

The sample consists of 449 individuals who are the citizens of one of the Polish cities, Bydgoszcz, about 350 000 citizens. In the studies the whole group of individuals is split to 160 employees (100 females and 60 males) and 289 retirees (186 females and 103 males). Additionally, the group of retirees is split to two subgroups: 106 students of U3A (79 females and 27 males) and 183 non-students of U3A (107 females and 76 males).

In the present work, we study the overall quality of life and general health. Figures 1-9 show maps obtained using the correspondence analysis. Figure 1 shows the degree of satisfaction on the overall quality of life and general health of particular groups of individuals with vocational education in a 3-point scale (*positive, negative, neutral*). Figures 2 and 3 present the same relations but for the individuals with different education: high school (Figure 2) and university (Figure 3). These relations are also shown in a 5-point scale (*A-1,A-2,... A-5*) for the individuals with the vocational education (Figure 4), with the high school (Figure 5) and with the university education (Figure 6). The number of individuals with doctor's degree is small and therefore the correspondence analysis can be applied only to the 3-point scale (Figure 7). Figures 8 and 9 show the degree of satisfaction on the overall quality of life and general health of particular groups of individuals with different job positions in the 3-point scale (Figure 8) and in the 5-point scale (Figure 9).

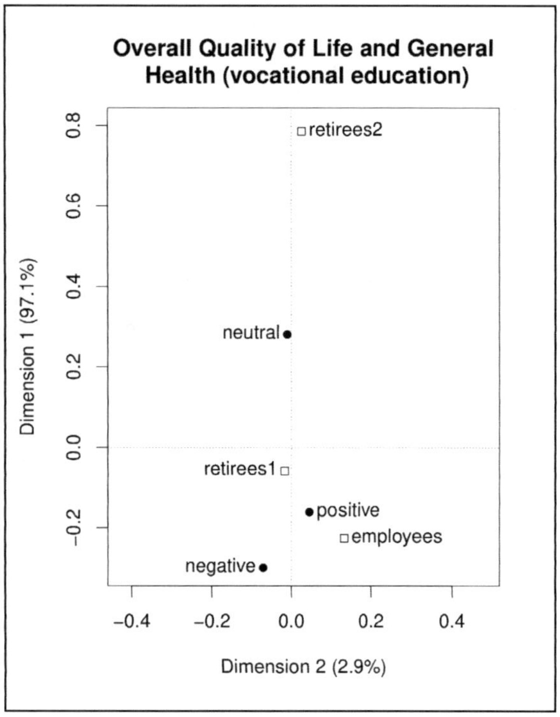

Figure 1. Map obtained using the correspondence analysis (rectangles denote groups of individuals with vocational education, circles denote answers in 3-point scale).

In all the figures, two kinds of points appear: empty rectangles and full circles. Empty rectangles denote three different subgroups of individuals: *employees*, *retirees1* (non-students of U3A), and *retirees2* (students of U3A). Full circles denote answers to the questions in the WHOQOL-BREF questionnaire. The answers are labeled as *A-1, A-2, ...A-5* in Figures 4, 5, 6, and 9, respectively. The respondents could choose between these five answers, where *A-1* corresponds to very poor quality of life and health and *A-5* to the very good ones. In order to consider the problem in a lower resolution we collect two negative answers, *A-1* and *A-2*, to one negative answer and two positive answers, *A-4* and *A-5*, – to one positive. In Figures 1, 2, 3, 7, 8 full circles denote the answers: *positive*, *neutral* (*A-3*), and *negative*.

Figures 1-7 show maps for the individuals with different kinds of education: vocational (Figures 1, 4), high school (Figures 2, 5), university (Figures 3, 6), doctor's degree (Figure 7).

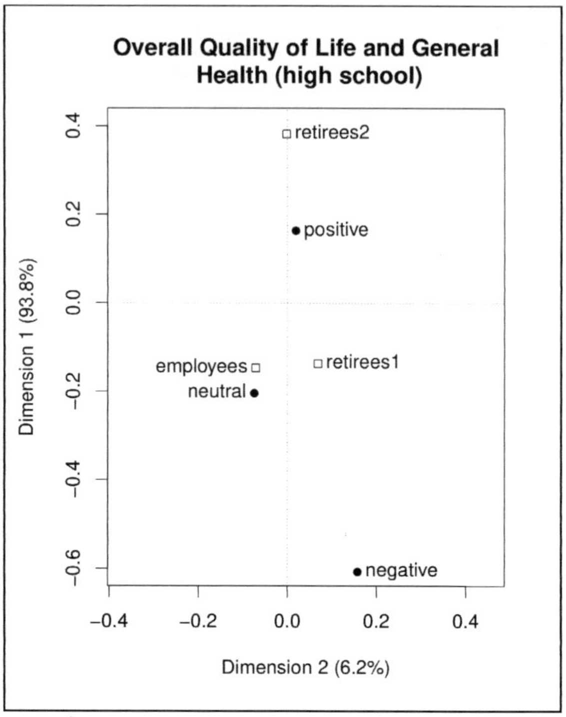

Figure 2. Map obtained using the correspondence analysis (rectangles denote groups of individuals with high school education, circles denote answers in 3-point scale).

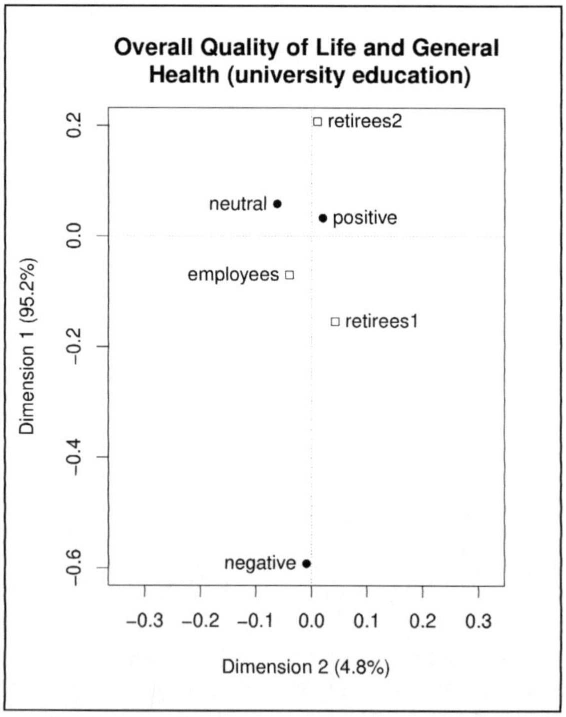

Figure 3. Map obtained using the correspondence analysis (rectangles denote groups of individuals with university education, circles denote answers in 3-point scale).

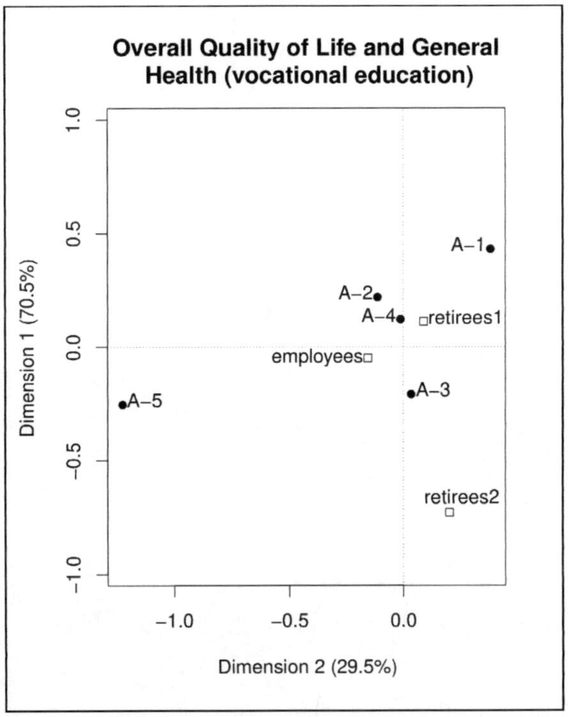

Figure 4. Map obtained using the correspondence analysis (rectangles denote groups of individuals with vocational education, circles denote answers in 5-point scale).

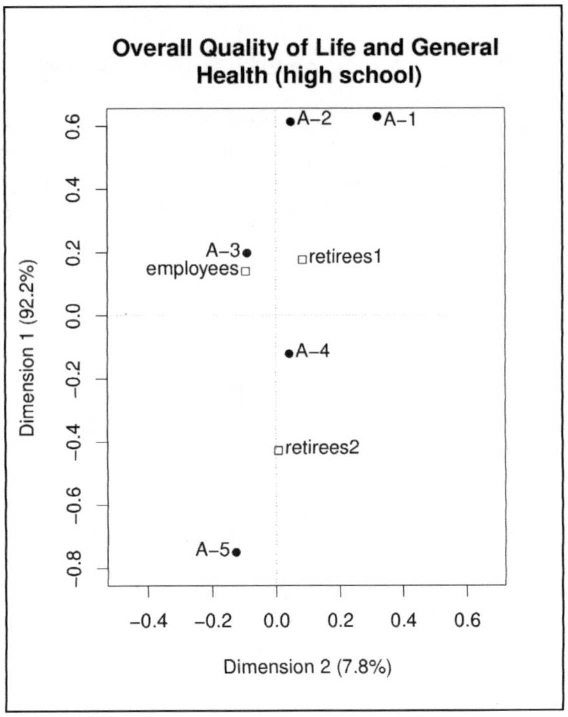

Figure 5. Map obtained using the correspondence analysis (rectangles denote groups of individuals with high school education, circles denote answers in 5-point scale).

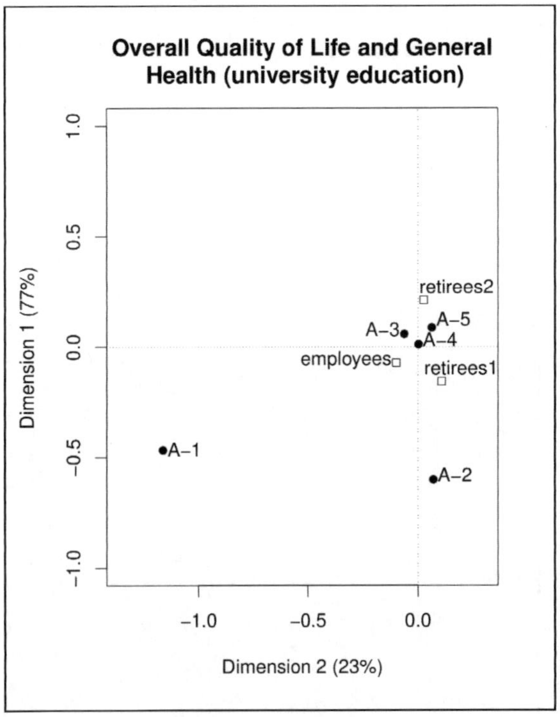

Figure 6. Map obtained using the correspondence analysis (rectangles denote groups of individuals with university education, circles denote answers in 5-point scale).

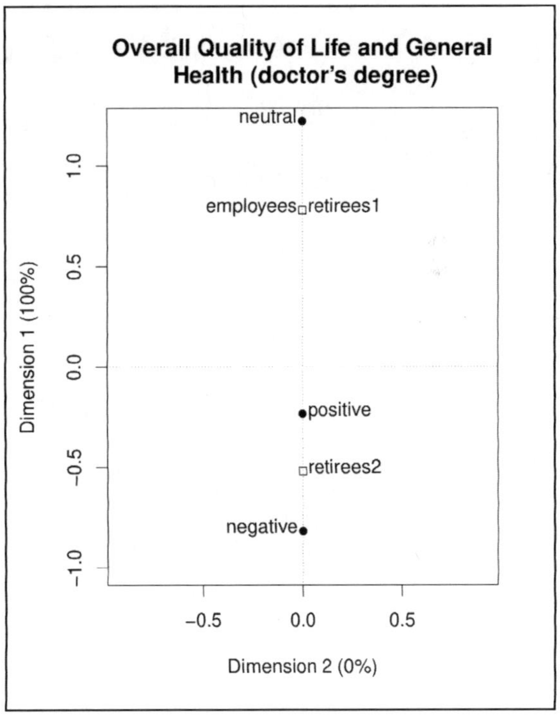

Figure 7. Map obtained using the correspondence analysis (rectangles denote groups of individuals with doctor's degree, circles denote answers in 3-point scale).

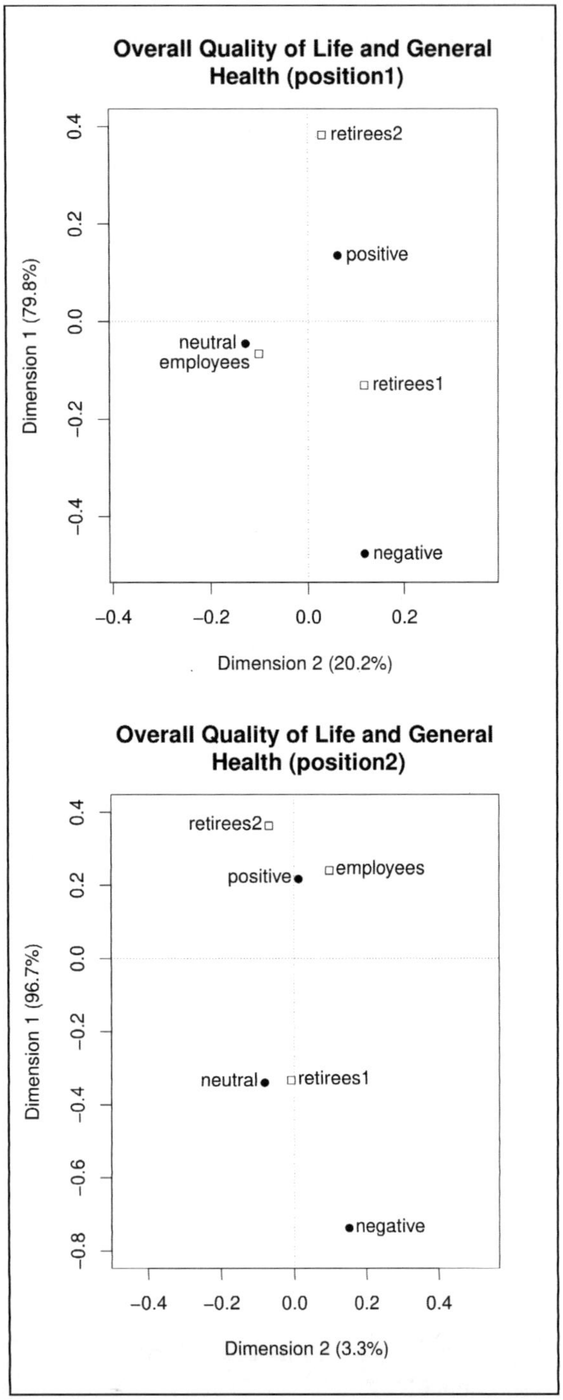

Figure 8. Map obtained using the correspondence analysis (rectangles denote groups of individuals with *position1* - top panel and with *position2* - bottom panel, circles denote answers in 3-point scale).

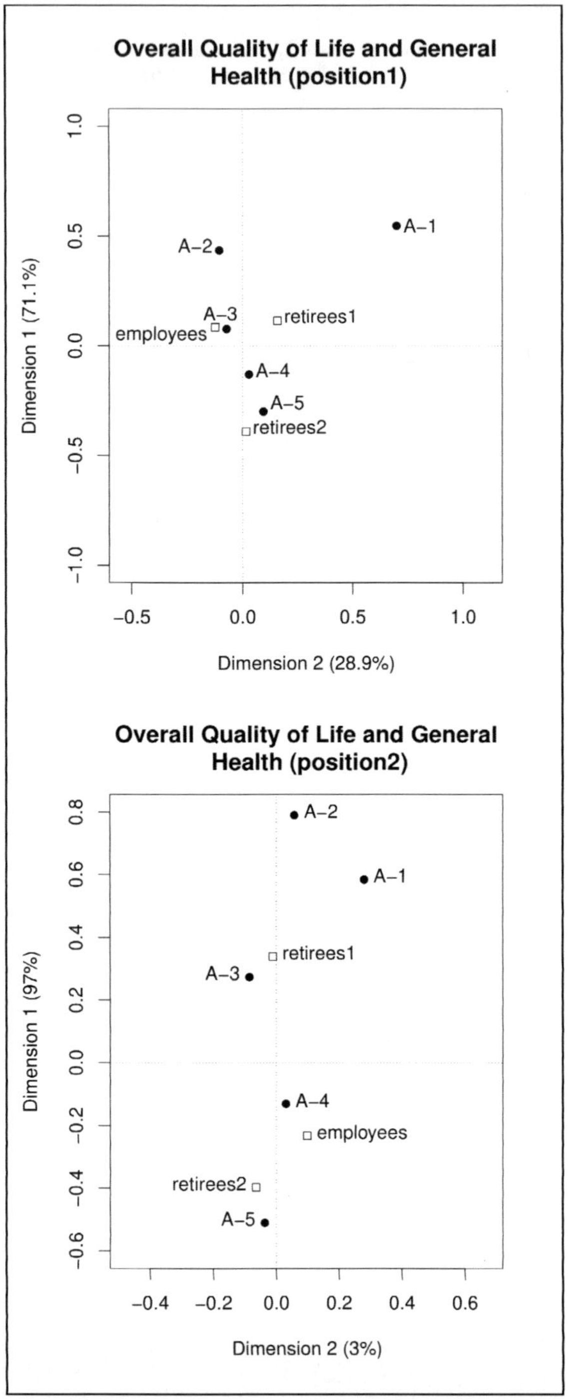

Figure 9. Map obtained using the correspondence analysis (rectangles denote groups of individuals with *position1* - top panel and with *position2* - bottom panel, circles denote answers in 5-point scale).

We also considered different job positions during the employment (Figures 8, 9): Staff is labeled as *position1*; supervisor/manager, director/president, and business owner are labeled as *position2*.

The satisfaction level on the overall quality of life and general health both for the group of employees and for the retirees strongly depends on the level of education: the structures of maps are different for the vocational, the high school, and for the university educations. Also, the intellectual activity during the retirement, such as attending the University of the Third Age, influences the degree of this satisfaction. We observe that in the maps, the points representing *retirees2* and *retirees1* are located rather far away from each other. In the cases of the dependence on the job position, the observations are similar. The structures of maps for *position1* and for *position2* are different. The job position is an important factor influencing the satisfaction level on the overall quality of life and general health. The point representing employees is close to the neutral answer for *position1* and close to the positive one for *position2* (Figure 8). In the higher resolution this is *A-3* and *A-4* respectively (Figure 9). The group of students of U3A (*retirees2*) is the most satisfied group both for *position1* and *position2*: the closest point to *A-5* is *retirees2*.

IV. Conclusion

Summarizing, we have shown that both education level and the job position are important factors influencing the change of the satisfaction level on the overall quality of life and general health related to the retirement threshold. We have also shown that the correspondence analysis is a convenient tool for such kind of studies.

The same methodology we are also going to apply for the studies on changes of the quality of life in different domains after the retirement threshold. Some preliminary results have already been obtained.

References

[1] A. Syse, M. Veenstra, T. Furunes, R. J. Mykletun, and P. E. Solem, "Changes in Health and Health Behavior Associated With Retirement", Journal of Aging and Health, vol. 29, pp. 99-127, 2017.

[2] M. Manty et al., "Pre-retirement physical working conditions and changes in physical health functioning during retirement transition process", Scandinavian Journal of Work, Environment & Health, vol. 42, pp. 405-412, 2016.

[3] B. Kail, "The Mental and Physical Health Consequences of Changes in Private Insurance Before and After Early Retirement", The Journals of Gerontology Series B: Psychological Sciences and Social Sciences, vol. 71, pp. 358-368, 2016.

[4] C. A. Maher et al., "Changes in use of time, activity patterns, and health and wellbeing across retirement: design and methods of the life after work study", BMC Public Health, vol. 13, pp. 952, 2013.

[5] A. Helldan, T. Lallukka, O. Rahkonen, and E. Lahelma, "Changes in healthy food habits after transition to old age retirement", European Journal of Public Health, vol. 22, pp. 582-586, 2012.

[6] R. Gueorguieva, J. L. Sindelar, R. Wu, and W. T. Gallo, "Differential changes in body mass index after retirement by occupation: hierarchical models", International Journal of Public Health, vol. 56, pp. 111-116, 2011.

[7] K. Regan et al., "Changes in Physical Activity and Function with Transition to Retirement Living: A Pilot Study", Canadian Journal on Aging / La Revue canadienne du vieillissement, vol. 35, pp. 526-532, 2016.

[8] D. Van Dyck, G. Cardon, and I. De Bourdeaudhuij, "Longitudinal changes in physical activity and sedentary time in adults around retirement age: what is the moderating role of retirement status, gender and educational level?", BMC Public Health, vol. 16, pp. 1125, 2016.

[9] J. Lahti, A. Holstila, M. Manty, E. Lahelma, and O. Rahkonen, "Changes in leisure time physical activity and subsequent disability retirement: A register-linked cohort study", International Journal of Behavioral Nutrition and Physical Activity, vol. 13, pp. 99, 2016.

[10] S. Stenholm et al., "Changes in physical activity during transition to retirement: a cohort study", International Journal of Behavioral Nutrition and Physical Activity, vol. 13, pp. 51, 2016.

[11] J. Sprod, K. Ferrar, T. Olds, and C. Maher "Changes in sedentary behaviours across the retirement transition: a systematic review", Age and Ageing, vol. 44, pp. 918-925, 2015.

[12] S. McDonald, N. O'Brien, M. White, and F. F. Sniehotta, "Changes in physical activity during the retirement transition: a theory-based, qualitative interview study", International Journal of Behavioral Nutrition and Physical Activity, vol. 12, pp. 25, 2015.

[13] M. Menai et al., "Changes in Sedentary Behaviours and Associations with Physical Activity through Retirement: A 6-Year Longitudinal Study", PLOS ONE, vol. 9, pp. e106850, 2014.

[14] I. Barnett, E. van Sluijs, D. Ogilvie, and N. J. Wareham, "Changes in household, transport and recreational physical activity and television viewing time across the transition to retirement: longitudinal evidence from the EPIC-Norfolk cohort", Journal of Epidemiology and Community Health, vol. 68, pp. 747-753, 2014.

[15] N. Sjosten et al., "Change in physical activity and weight in relation to retirement: the French GAZEL Cohort Study", BMJ Open, vol. 2, pp. e000522, 2012.

[16] J. Lahti, M. Laaksonen, E. Lahelma, and O. Rahkonen, "Changes in leisure-time physical activity after transition to retirement: a follow-up study", International Journal of Behavioral Nutrition and Physical Activity, vol. 8, pp. 36, 2011.

[17] M. Touvier et al., "Changes in leisure-time physical activity and sedentary behaviour at retirement: a prospective study in middle-aged French subjects", International Journal of Behavioral Nutrition and Physical Activity, vol. 7, pp. 14, 2010.

[18] T.S. Olds et al., "Everybody's working for the weekend: changes in enjoyment of everyday activities across the retirement threshold", Age and Ageing, vol. 45, pp. 850-855, 2016.

[19] E. W. Hagen, J. H. Barnet, L. Hale, and P. E. Peppard, "Changes in Sleep Duration and Sleep Timing Associated with Retirement Transitions", leep, vol. 39, pp. 665-673, 2016.

[20] J. Berg, "Mobility changes during the first years of retirement. Quality in Ageing and Older Adults", vol. 17, pp. 131-140, 2016.

[21] M. N. Vercambre, O. I. Okereke, I. Kawachi, F. Grodstein, and J. H. Kang, "Self-Reported Change in Quality of Life with Retirement and Later Cognitive Decline: Prospective Data from the Nurses' Health Study", Journal of Alzheimer's Disease, vol. 52, pp. 887-897, 2016.

[22] G. F. Barrett and M. Kecmanovic, "Changes in subjective well-being with retirement: assessing savings adequacy",Applied Economics, vol. 45, pp. 4883-4893, 2013.

[23] P. C. Castro et al., "Influence of the Open University for the Third Age (UATI) and the revitalization program (REVT) on quality of life in middle-aged and elderly adults", Revista Brasileira de Fisioterapia, vol. 11, pp. 461-467, 2007.

[24] L. Hebestreit, "The role of the University of the Third Age in meeting needs of adult learners in Victoria, Australia", Australian Journal of Adult Learning, vol. 48, pp. 547-565, 2008.

[25] J. G. Sonati et al., "Body composition and quality of life (QoL) of the elderly offered by the "University Third Age" (UTA) in Brazil", Archives of Gerontology and Geriatrics, vol. 52, pp. e31-e35, 2011.

[26] H. Zielińska-Więczkowska, K. Kędziora-Kornatowska, and W. Ciemnoczołowski, "Evaluation of quality of life (QoL) of students of the University of Third Age (U3A) on the basis of socio-demographic factors and health status", Archives of Gerontology and Geriatrics, vol. 53, pp. e198-e202, 2011.

[27] J. Krzepota, E. Biernat, and B. Florkiewicz, "The relationship between levels of physical activity and quality of life among students of the University of the Third Age", Central European Journal of Public Health, vol. 23, pp. 335-339, 2015.

[28] J. Mackowicz and J. Wnek-Gozdek, "It's never too late to learn"-How does the Polish U3A change the quality of life for seniors?" Educational Gerontology, vol. 42, pp. 186-197, 2016.

[29] L. Wołowicka, "Jakość życia w naukach medycznych", Wydawnictwo Uczelniane AM w Poznaniu, Poznań, pp. 231-281, 2001.

[30] H. O. Hirschfeld, "A connection between correlation and contingency", Proceedings of the Cambridge Philosophical Society, vol. 31, pp. 520-524, 1935.

[31] P. Wąż and D. Bielińska-Wąż, "Moments of Inertia of Spectra and Distribution Moments as Molecular Descriptors", MATCH Communications in Mathematical and in Computer Chemistry, vol. 70, pp. 851-865, 2013.

[32] P. Wąż and D. Bielińska-Wąż, "3D-dynamic representation of DNA sequences", Journal of Molecular Modeling, vol. 20, pp. 2141, 2014.

Influence of the Education Level on Health of Elderly People

Agnieszka Bielińska*, Mikołaj Majkowicz|| **, Dorota Bielińska-Wąż§ and Piotr Wąż¶
* Department of Quality of Life Research, Medical University of Gdańsk
Gdańsk, Poland
Email: agnieszka.bielinska1984@gmail.com
|| Department of Quality of Life Research, Medical University of Gdańsk
Gdańsk, Poland
Email: mmajk@gumed.edu.pl
** Department of Public Health, Pomeranian University in Słupsk
Słupsk, Poland
§ Department of Radiological Informatics and Statistics, Medical University of Gdańsk
Gdańsk, Poland
Email: djwaz@gumed.edu.pl
¶ Department of Nuclear Medicine, Medical University of Gdańsk
Gdańsk, Poland
Email: phwaz@gumed.edu.pl

Abstract—**In the present work, the relation between the level of education of elderly people and their health and quality of life has been studied. The studies have been performed in Poland. The collected data have been analyzed using the World Health Organization Quality of Life-BREF (WHOQOL-BREF) questionnaire (an abbreviated version of the WHOQOL-100), and the Geriatric Depression Scale. In both cases, we have observed a large influence of the education level on health. In most cases, the higher the education, the more positive the results. A large influence of the marital status on the overall quality of life and general health of the Polish pensioners has also been observed.**

Keywords–Social medicine; Biomedical datasets; Medical informatics; Visualizing data; Correspondence analysis; World Health Organization Quality of Life-BREF (WHOQOL-BREF); Geriatric Depression Scale (GDS)

I. INTRODUCTION

Naturally, the human health deteriorates with passing time. What is the influence of different factors on this natural process? For instance, Cho et al. consider such factors as education and past life experiences on successful aging and subjective well-being among oldest-old adults [1]. Johnson et al. study the influence of education and personality on health in the oldest old [2]. The authors also consider other factors, for example physical activity [3][4] or the attendance of lectures at the University of the Third Age [5]-[8], which help in successful aging.

In the present work, we study the influence of the education level and the marital status of elderly people on the overall quality of life and general health. We also examine the influence of the education on the level of depression. The studies have been performed on a group of retired residents of Poland.

In Section II we describe the methods. The results in the form of maps obtained using the correspondence analysis and the spine plots are shown in Section III. Section IV summarizes the results.

II. METHODS

The studies were performed from February 2017 to May 2017 in Bydgoszcz, the eighth largest city in Poland, about 350 000 citizens. The studied group consists of 289 retirees (186 females and 103 males). All the participants were evaluated using the Polish version of the World Health Organization Quality of Life-BREF (WHOQOL-BREF) questionnaire [9][10] and using the Geriatric Depression Scale (GDS) [11]. They could declare their education as: elementary school, vocational education, high school, or university education. We applied the correspondence analysis for the graphical representation of the results. Originally, this method was introduced by Hirschfeld [12]. Later, the method has been improved by many authors [13][14]. In these methods, one creates maps in which the objects under consideration are represented by points distributed in a specific way. Distances between these points are related to properly defined similarity classes of the objects. Thus, by studying the distribution of the distances one can classify the objects according to some selected similarity criteria. This methodology may be applied in many areas of science. In particular, we have recently created an analog of this kind of approach in bioinformatics [15].

In this work, the points on the maps represent different groups of individuals and their answers to questions. In order to classify the set of studied individuals, we analyze the distribution of the distances between the points representing the answers and the points representing groups of individuals.

III. RESULTS AND DISCUSSION

Figures 1, 2 and 3 show 2D-maps derived from the correspondence analysis. They show the relations between the general satisfaction of the retirees with the quality of life and health and their age, marital status, and education, respectively.

There are two different kinds of points denoted in the maps. Empty rectangles denote groups of individuals. In Figure 1, empty rectangles correspond to the groups of individuals of different age: 51-64; 65-79, and above 80. In Figure 2, these symbols correspond to the groups with different marital status:

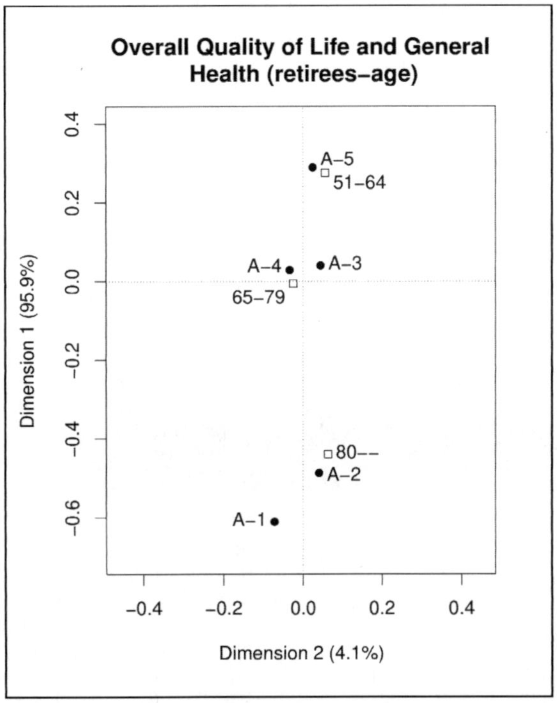

Figure 1. 2D-map obtained using the correspondence analysis method (circles denote answers, rectangles denote groups of individuals of different age).

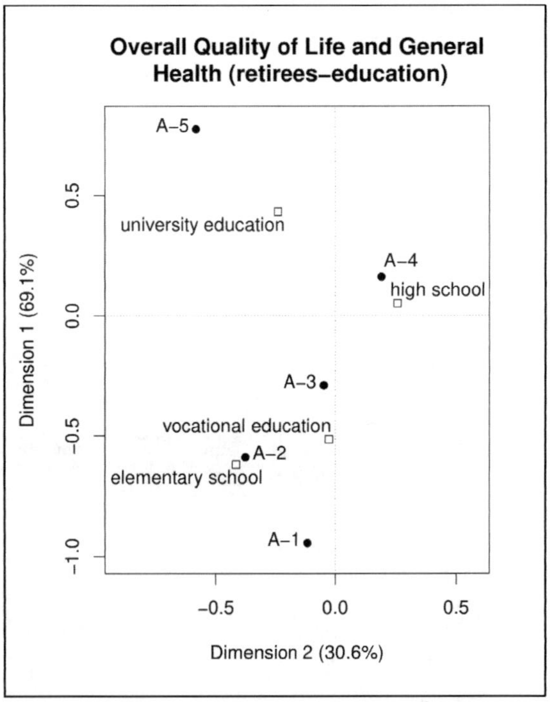

Figure 3. 2D-map obtained using the correspondence analysis method (circles denote answers, rectangles denote groups of individuals with different education).

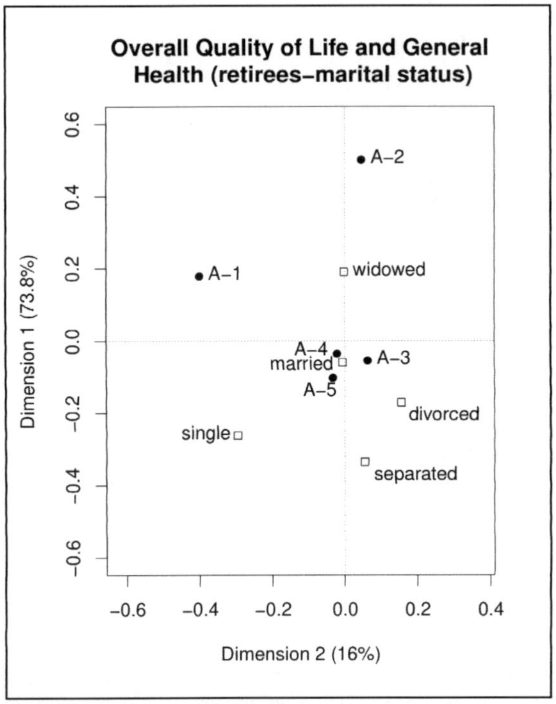

Figure 2. 2D-map obtained using the correspondence analysis method (circles denote answers, rectangles denote groups of individuals with different marital status).

single, married, widowed, divorced, separated. In Figure 3, the groups of individuals with different education are considered: elementary school, vocational education, high school, and university education. Full circles denote answers to the questions in the WHOQOL-BREF questionnaire about overall quality of life and general health. One can choose between five answers denoted in the figure as *A-1, A-2,... A-5*. *A-1* corresponds to the least favorable quality of life and *A-5* to the most favorable one.

As expected, the closest point to *A-5* is the youngest group *51-64*. The oldest group *80 - –* estimates its overall quality of life and general health as the worst compared to other groups (Figure 1).

Considering the marital status, the group *married* estimates its overall quality of life as the best compared to other groups and *widowed* as the worst (Figure 2).

Considering the education, the closest point to *A-5* is *university education*. It means that this group of individuals estimates their overall quality of life and general health as the best compared to other groups. The closest point to *A-1* is *elementary school* (Figure 3).

Figures 4, 5 and 6 show the results obtained using the Geriatric Depression Scale.

We used the 15-item version of GDS. The questions are the following:

1) Are you basically satisfied with your life? yes / **no**
2) Have you dropped many of your activities and interests? **yes** / no
3) Do you feel that your life is empty? **yes** / no

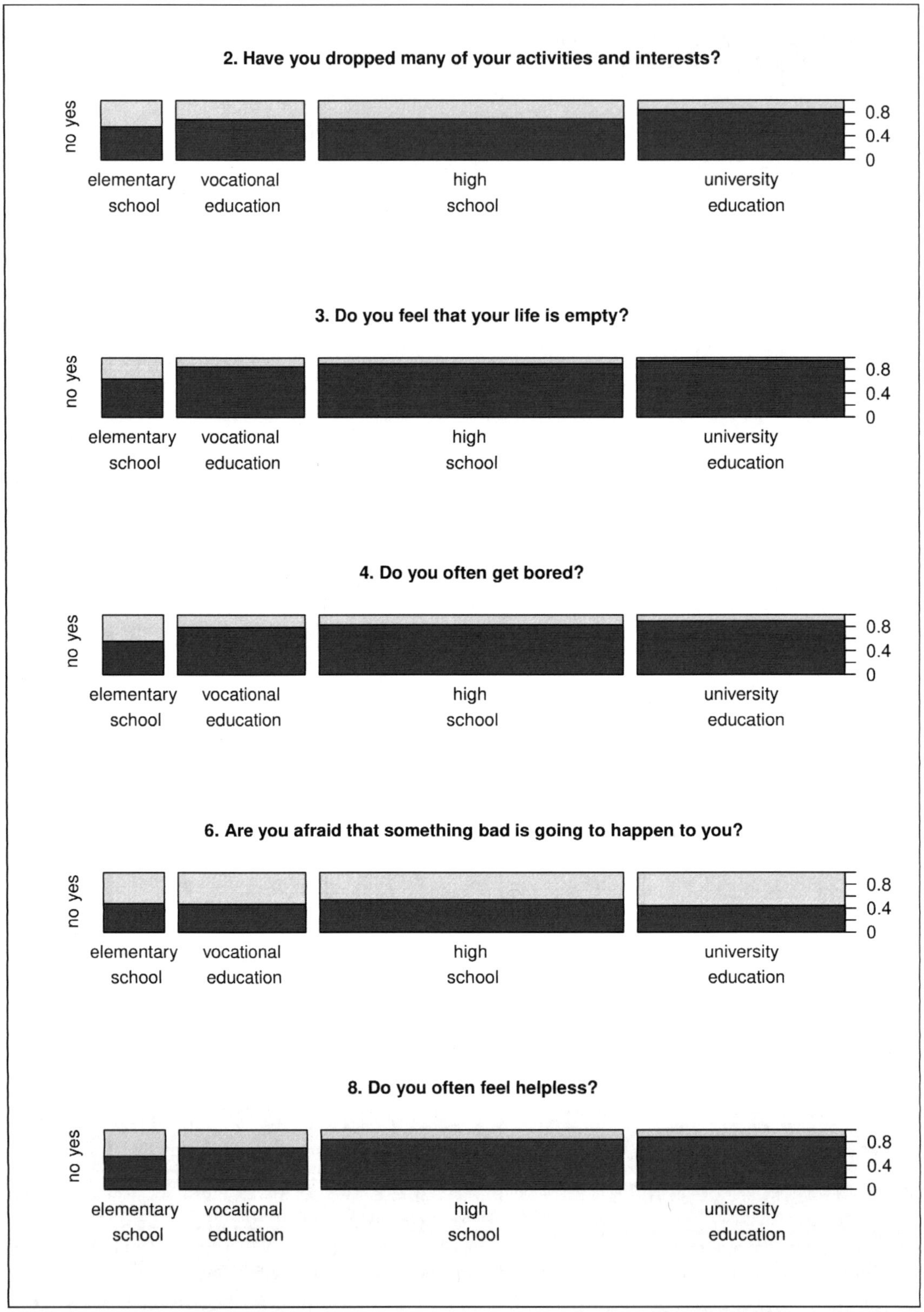

Figure 4. Spine-plots showing the dependence of the percentage of positive and negative answers to a particular question on the education level of the respondents (questions No. 2,3,4,6,8 for which answer 'yes' indicates depressive symptoms).

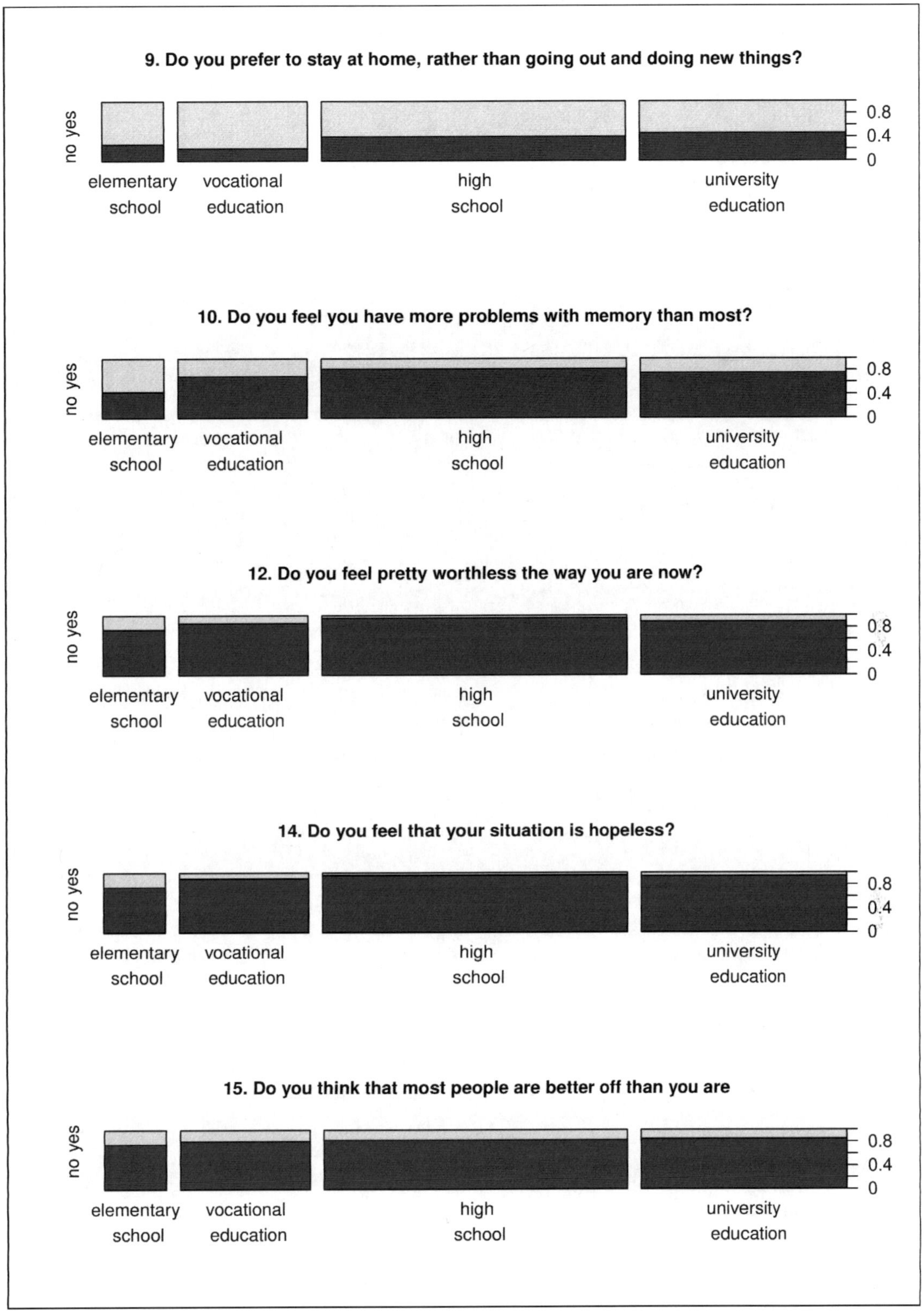

Figure 5. Spine-plots showing the dependence of the percentage of positive and negative answers to a particular question on the education level of the respondents (questions No. 9,10,12,14,15 for which answer 'yes' indicates depressive symptoms).

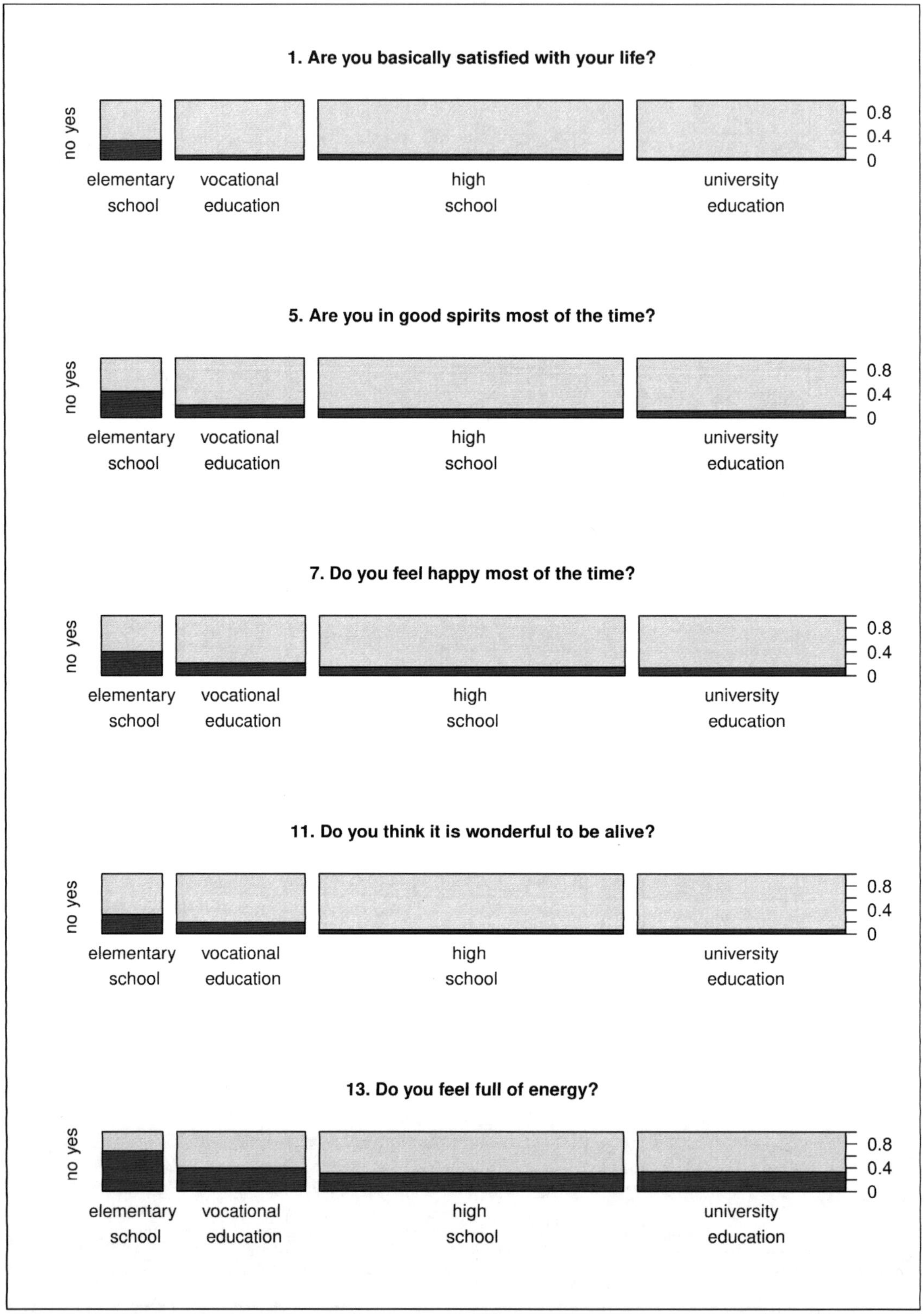

Figure 6. Spine-plots showing the dependence of the percentage of positive and negative answers to a particular question on the education level of the respondents (questions for which answer 'no' indicates depressive symptoms).

4) Do you often get bored? **yes** / no
5) Are you in good spirits most of the time? yes / **no**
6) Are you afraid that something bad is going to happen to you? **yes** / no
7) Do you feel happy most of the time? yes / **no**
8) Do you often feel helpless? **yes** / no
9) Do you prefer to stay at home, rather than going out and doing new things? **yes** / no
10) Do you feel you have more problems with memory than most? **yes** / no
11) Do you think it is wonderful to be alive? yes / **no**
12) Do you feel pretty worthless the way you are now? **yes** / no
13) Do you feel full of energy? yes / **no**
14) Do you feel that your situation is hopeless? **yes** / no
15) Do you think that most people are better off than you are? **yes** / no

Answers in bold indicate depressive symptoms. Answer **yes** means depressiveness in questions No: 2,3,4,6,8,9,10,12,14,15 (Figures 4 and 5). Answer **no** means depressiveness in all other cases, i.e., questions No: 1,5,7,11,13 (Figure 6). For each bolded answer is given 1 point.

A score greater than 5 points suggests mild depressive symptoms. If the score is greater than 10 points it may be depressive disorder. Such individual should be tested for depression.

Figures 4, 5 and 6 show the spine-plots. The width is the largest for *high school* so the largest number of individuals declare this kind of education. The smallest number of individuals declare *elementary school* education (the smallest width in the spine-plots). Dark color represents the answer no. In all the questions in which answer 'no' means depression (1,5,7,11,13) the percentage of answers 'no' is the largest for individuals having elementary education compared to other groups (Figure 6). Analogously, in other questions in which answer 'yes' indicates depression 2,3,4,8,10,12,14,15 also the largest percentage of answers 'yes' is for individuals with elementary education compared to other groups (Figures 4 and 5). The exceptions are only two questions: No. 6 and 9. For these questions, the answer 'yes' indicates depressive symptoms. In question No. 6 all the groups answer approximately in a similar way. In question 9 the differences between groups are also not large but *vocational education* has slightly larger percentage of 'depressive' answers (yes) comparing to other groups.

IV. Conclusion

Summarizing, the education and the marital status strongly influence the overall quality of life and general health. We have also shown that among Polish elders, the depressive symptoms are related to the education. Higher education represents an important positive factor affecting this problem. Generally, the higher the education, the better the overall quality of life and health.

In future work, we will concentrate on studies related to the influence of factors other than education on depressive symptoms. In particular, we will check the significance of sex and marital status.

References

[1] J. Cho, P. Martin, and L. W. Poon, "Successful Aging and Subjective Well-Being Among Oldest-Old Adults", Gerontologist vol. 55, pp. 132-143, 2015.

[2] M. C. Johnson, N. Budd, P. Martin, T. Schofield, and L. Poon, "The influence of education and personality on health in the oldest old: Findings from the Georgia Centenarian study", Gerontologist vol. 53, pp. 475-476, Supplement: 1 , 2013.

[3] M. Choi, M. Lee, M. J. Lee, and D. Jung, "Physical activity, quality of life and successful ageing among community-dwelling older adults", International Nursing Review vol. 64, pp. 396-404, 2017.

[4] K. Conner et al., "Practical applications of physical activity for successful cognitive aging", Jaapa-Journal of the American Academy of Physician Assistants vol. 30, pp. 30-35, 2017.

[5] P. C Castro et al., "Influence of the Open University for the Third Age (UATI) and the revitalization program (REVT) on quality of life in middle-aged and elderly adults", Revista Brasileira de Fisioterapia vol. 11, pp. 461-467, 2007.

[6] L. Hebestreit, "The role of the University of the Third Age in meeting needs of adult learners in Victoria, Australia", Australian Journal of Adult Learning vol. 48, pp. 547-565, 2008.

[7] J. G. Sonati et al., "Body composition and quality of life (QoL) of the elderly offered by the "University Third Age" (UTA) in Brazil", Archives of Gerontology and Geriatrics vol. 52, pp. e31-e35, 2011.

[8] J. Mackowicz and J. Wnek-Gozdek, "It's never too late to learn"-How does the Polish U3A change the quality of life for seniors?", Educational Gerontology vol. 42 pp. 186-197, 2016.

[9] http://www.who.int/substance_abuse/research_tools/whoqolbref/en/

[10] L. Wołowicka, "Quality of life in medical sciences", Wydawnictwo Uczelniane AM w Poznaniu, Poznań, pp. 231-281, 2001.

[11] J. I. Sheikh and J. A. Yesavage, Geriatric Depression Scale (GDS): Recent evidence and development of a shorter version, Clinical Gerontologist vol. 5, pp. 165-173, 1986.

[12] H. O. Hirschfeld, "A connection between correlation and contingency", Proceedings of the Cambridge Philosophical Society vol. 31, pp. 520-524, 1935.

[13] J.-P. Benzécri, "L'Analyse des Données", L'Analyse des Correspondances, Paris, France, Dunod, vol. II. 1973.

[14] M. Greenacre, "Correspondence Analysis in Practice", Second Edition, London, Chapman & Hall/CRC, 2007.

[15] P. Wąż, D. Bielińska-Wąż, and A. Nandy, "Descriptors of 2D-dynamic graphs as a classification tool of DNA sequences", Journal of Mathematical Chemistry vol. 52, pp. 132-140, 2014.

eCare Tool for Person-Centred Care of People with Dementia in Nursing Homes

Kelber Pieter[1], De Smet Herwig[2], Van Rompaey Michèle[1], Kerstens Birgit[1]
Dept. of Nursing[1] & Dept. of Applied Informatics[2]
Karel de Grote University College
Brusselstraat 45, 2018 Antwerp, Belgium
emails: pieter.kelber@kdg.be, herwig.desmet@kdg.be, michele.vanrompaey@kdg.be, birgit.kerstens@kdg.be

Van Bogaert Peter[3], Franck Erik[1,3]
Dept. of Nursing and Midwifery Sciences
Centre for Research in Innovation and Care[3]
University of Antwerp
Prinsstraat 13, 2000 Antwerp, Belgium
emails: peter.vanbogaert@uantwerpen.be, erik.franck@uantwerpen.be

Abstract—**Electronic tools have the potential to support healthcare staff in providing more person-centred care to persons living with dementia in residential nursing homes. The paper describes work in progress to develop an eCare tool that provides relevant information to caregivers about the needs of persons living with dementia and appropriate actions related to behavioural and psychological symptoms of dementia. The eCare tool, an eHealth technology to be integrated in electronic care records kept in nursing homes, is particularly relevant in the context of 'nurse team applications'. Increased person-centred care is linked to increase in job satisfaction and personal competence among nursing staff, resulting in lower turnover figures in nursing homes.**

Keywords-eCare tool; person-centred care; dementia; life stories; behavioural and psychological symptoms of dementia.

I. INTRODUCTION

The World Health Organization (WHO) estimates that 50 million people were living with dementia worldwide in 2015. The total number of Persons Living With Dementia (PLWD) is projected to increase to 80 million in 2030 and to triple to 152 million in 2050 [1]. Dementia affects cognitive functions, often followed by Behavioural and Psychological Symptoms of Dementia (BPSD), such as physical and verbal aggression, wandering, calling out; agitation and anxiety, hallucinations; depression and apathy [2]. Currently, there is no curative treatment available.

Dementia progressively inhibits the affected person's ability to communicate and satisfy basic human needs. PLWD often express or manifest their unmet needs through socially inappropriate or unusual behaviour including the above symptoms [3], which may lead to negative feelings in professional caregivers and other nursing home residents [4]. When caregivers focus on the physical needs, this can come at the expense of the psychosocial needs of this person. This incorporates an actual risk that nursing care tends to be more task oriented, objectifying and even depersonalizing. Some caregivers believe that progression to severe dementia leads to the gradual loss of personhood until there is nothing left of the person. It poses a real threat that the life of this PLWD is perceived to be meaningless, thus the role of the caregiver loses meaning too [5].

Already 20 years ago Kitwood introduced the conceptual model of Person-Centred Care (PCC) based on the idea that the personhood of PLWD is not lost, but is concealed [6]. Whether the psychological needs of PLWD are met and if personhood is maintained or strengthened, is largely determined by caregiver behaviour and environment [7][8].

Recent systematic reviews demonstrate the effectiveness of non-pharmacological interventions of PCC to reduce BPSD [9]-[12]. Reminiscence therapy (reliving experiences from the past by talking about recognizable photos, objects, etc.) [13]-[16], music therapy [17]-[19] and the use of biographic information of the PLWD during communication or care activities [20][21] are successful interventions that reduce BPSD.

PCC also positively effects the professional caregivers' personal accomplishment and job satisfaction [22]: e.g., reminiscence improves caregivers' communication with residents and increases their knowledge of the resident's background significantly [15][16], music therapy enhances caregiving techniques [18], while life story work enables healthcare staff "to see the person behind the patient" and empowers them to engage in "genuine participatory practices" [23][24]. There is also strong evidence that professional caregivers benefit from learning and skill development in PCC interventions and communication in order to sustain the effects of those interventions to apply person-centred care into practice [9][25].

Lastly, the above positive effects influence the perception of quality of care provided by the nursing home: the person-centred care staff feels more satisfied and competent to perform their job while the PLWD and their relatives feel more satisfied with the care received [25].

We are currently exploring the feasibility of an electronic tool that guides healthcare staff to integrate person-centred care in the daily care of persons living with dementia in residential nursing homes. The results of the feasibility study are expected to feed into a two-year implementation study.

Section II describes the objectives, study design and the work performed so far. Section III presents the preliminary findings, while Section IV elaborates on the work in progress for the development of the (e) learning and training modules and the eCare tool into the existing care records. Section V holds the conclusion and future work.

II. STUDY CHARACTERISTICS

A. Objective

Our study aims to develop and validate a practical electronic tool that enables healthcare providers to integrate person-centred care in the daily practice. The current study funding supports the preparatory work to test the feasibility of the tool, to identify the interventions preferred by healthcare staff to provide PCC to persons living with dementia and to outline the functional specifications of the eCare tool that facilitates PCC in nursing homes.

B. Study design

We conducted a literature review on the effectiveness of non-pharmacological interventions and person-centred care for PLWD. We held two focus group discussions with professional caregivers working in residential nursing homes to identify which PCC interventions they prioritize to be integrated in the eCare tool. They were also asked to express the desired functional requirements of the eCare tool in terms of content and technology, and how it can be integrated in existing healthcare records. We conducted interviews to explore the market of software applications for PLWD and their caregivers. We assessed the economic and societal relevance of electronic tools for PCC interventions.

III. PRELIMINARY FINDINGS

A. Preliminary lessons learnt

Ongoing research about integration of PCC in the daily care of PLWD and our study results so far have given us insights into:

- the PCC interventions considered by caregivers as being the most relevant to address the (unmet) needs of PLWD, in particular life stories, reminiscence and music;
- the need for staff training in person-centred care and supportive actions to address and prevent BPSD;
- the preference/need to integrate the eCare tool in the existing electronic care records used in nursing homes.

B. Development of an eCare tool supporting PCC of PLWD

The focus groups with potential end users (i.e., professional caregivers of PLWD in nursing homes) discussed primarily the most effective interventions to guarantee PCC of PLWD, and the preferred PCC interventions to be integrated in the electronic care records. Coding and analysis of the focus group discussions in NVivo revealed that nursing staff prefers to incorporate two main instruments into the eCare tool: (1) 'know the person living with dementia' and (2) incidents related to BPSD. As mentioned above professional caregivers identify reminiscence therapy or use of life stories as effective interventions to get to know the caretaker better, but often they don't hold all relevant personal or biographic information at hand: the focus group participants clearly expressed interest in having this information readily available, for use in daily care or when observing BPSD.

Figure 1 reflects how these two elements would be included in the eCare tool whereby the tool on the one hand provides information to get to know the PLWD better and on the other hand hints to address and prevent agitation and other BPSD.

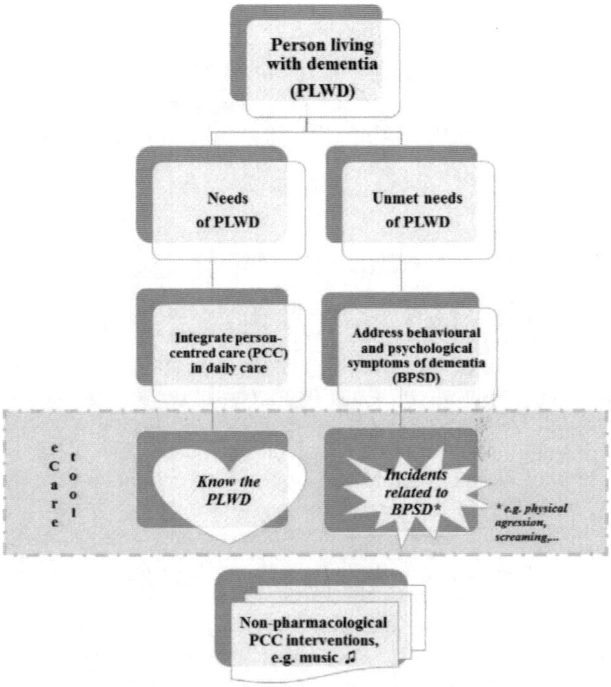

Figure 1. Elements to be integrated in the eCare tool.

1) Know the PLWD:

Nursing staff gives considerable importance to knowing the *'needs'* of the person taking care of. → The eCare tool should allow to *'know the PLWD'* through life story material and additional information received from the PLWD, their relatives and the nursing staff.

2) Incidents:

Nursing staff acknowledges the challenges to address *BPSD* in the most effective way. → The eCare tool should allow to register *incidents related to BPSD,* to analyse the incident and to suggest supportive and preventive actions.

IV. WORK IN PROGRESS

With the lessons learnt so far, we have prepared a study design to develop, test and evaluate the eCare tool. We will do this together with the end users, i.e., professional caregivers of PLWD in nursing home settings in Flanders. We are currently awaiting approval of study funding by the Flemish Agency for Innovation & Entrepreneurship and hope to start the study in September 2018.

A. Design of the eCare tool

In developing the eCare tool, we will follow the standard agile development cycle as depicted in Figure 2.

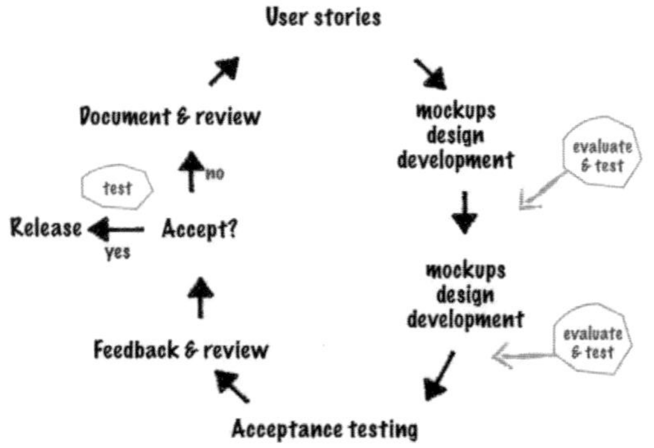

Figure 2. Agile development.

Before development can start, the requirements have to be analysed and described. Each requirement is split up in a number of user stories. A user story is an informal, natural language description of one or more features of a software system. User stories are often written from the perspective of an end user or user of a system. Each user story contains one well defined feature as in: *As a user I want to add an incident concerning this person*.

After this analysis and description phase, agile development is used to implement the software [26]. Agile is an iterative process. In each iteration, a prototype of the eCare tool is developed and tested. When a prototype is demo-ready, the end user can try it and give feedback, both at the level of the specifications and the user interface.

Agile uses test-driven development, meaning that testing is an essential part of the process [27]. Testing is performed at three levels. Each user story is tested in isolation (Unit Testing). After a set of units belonging to a specific subpart of the application has been finished, the subpart as a whole is tested (Integration Testing). When the total application is ready for testing, tests can be performed at that level (System Testing). We will also perform usability testing with the end users. This type of testing results in useful information about how easy it is to work with the software, and how good it looks. The eCare tool can be used a stand-alone mobile application but is supposed to be further integrated in the electronic care records used in nursing homes.

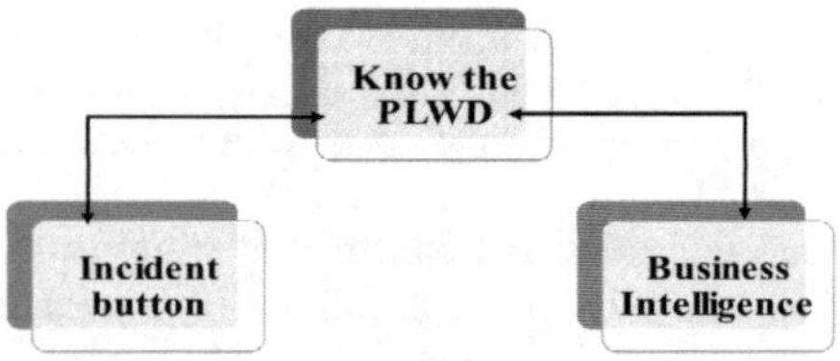

Figure 3. Application structure woth three modules.

We will focus on 3 software modules (Figure 3):

- Know the PLWD Module. For each PLWD, facts of life (incl. personal interests, needs and preferences) will be collected through intake interviews in the nursing homes and elements provided by the PLWD, their family and formal and informal caregivers. Both word processing software and database systems can be used for this purpose. Using Information Retrieval technology, relevant information is extracted from these texts. A feedback loop ensures the dynamically-evolving nature of the personal information and records the response of the person with dementia to certain interventions. This module is the central data repository for the other modules.
- Incident Button Module. A caregiver can use this module to register incidents related to BPSD. The eCare tool asks targeted questions to report BPSD, which can be answered quickly via a drop-down list of pre-selected options. A requirements analysis of BPSD follows by: 1) objectively describing the BPSD (based on pre-defined categories); excluding possible medical causes such as pain, urinary tract infection, constipation...; examining the context of behavioural change; 2) discussing and analysing the case at the next team meeting; 3) drawing up an action plan in team with focus on person-centred care and approach (first choice is a non-pharmacological intervention, followed by medical treatment if needed).
- Business Intelligence (BI) Module. Historical data regarding the incidents can be consulted by the head nurse or by other management staff, both about a specific individual and at the level of a unit. Appropriate visualizations help to analyse and understand the trends. Notifications are stored in the database in a structured way. This makes it possible to map out the historical course of a resident's BPSD over time. 'Drill down' techniques allow the user to request more details about the incidents themselves.

B. Design of the learning and training modules

The implementation of the eCare tool requires training in the correct use of PCC interventions and the eCare tool itself by the end users (i.e., professional caregivers of PLWD in nursing homes). Currently, we are testing a training programme with practical workshops on (1) how to address and prevent BPSD (e.g., physical aggression, screaming, etc.) and (2) how to integrate PCC in the daily practice through PCC interventions, i.e., reminiscence therapy, the use of music and the use of biographic information of the PLWD during communication or care activities. We are investigating the possibilities to provide the training programme in e-learning modules.

Once the eCare tool is ready to be integrated in the current (electronic) care records an additional training programme will be developed to train the end users on how to use the new applications/software modules.

V. CONCLUSION AND FUTURE WORK

The eCare tool is eHealth technology to be integrated in electronic care records kept in nursing homes and provides support to their healthcare staff for a person-centred approach of persons living with dementia. Increased quality of care and person-centred care are linked to increase in job satisfaction and personal competence among nursing staff, resulting in lower turnover figures in nursing homes.

The project team is currently conducting further discussions with IT experts and focus group discussions with healthcare providers and health managers about the functional requirements of the eCare tool. Future work depends on approval of study funding for two more years.

ETHICAL CONSIDERATIONS

Ethical approval will be requested from the Ethics Committee of the Antwerp University Hospital if application for further funding for development is successful.

ACKNOWLEDGMENTS

This study is funded by the Flemish Agency for Innovation & Entrepreneurship for technology transfer (TETRA) and co-financed by two Flemish networks of residential nursing homes (i.e., Curando and GZA Zorg en Wonen), Centre of Expertise Dementia Flanders and BlueHealth Innovation Center - supported by Microsoft.

REFERENCES

[1] WHO, Dementia - A public health priority (Infographic), 2017.

[2] S. Macfarlane and D. O'Connor, "Managing behavioural and psychological symptoms in dementia," Aust Prescr, vol. 39, no. 4, pp. 123-125, 2016.

[3] J. Cohen-Mansfield, "Nonpharmacologic treatment of behavioral disorders in dementia," Curr Treat Options Neurol, vol. 15, no. 6, pp. 765-85, 2013.

[4] A. Holst and L. Skar, "Formal caregivers' experiences of aggressive behaviour in older people living with dementia in nursing homes: A systematic review," Int J Older People Nurs, vol 12, no 4, pp. 1-12, 2017.

[5] D. Edvardsson, B. Winblad, and P. O. Sandman, "Person-centred care of people with severe Alzheimer's disease: current status and ways forward," Lancet Neurol, 2008. vol. 7, no. 4, pp. 362-7, 2008.

[6] T. Kitwood, "The experience of dementia," Aging & Mental Health, vol. 1, no. 1, pp. 13-22, 1997.

[7] G. Mitchell and J. Agnelli, "Person-centred care for people with dementia: Kitwood reconsidered," Nurs Stand, vol. 30, no. 7, pp. 46-50, 2015.

[8] D. Brooker, "What is person-centred care in dementia?"Reviews in Clinical Gerontology, vol. 13, pp. 215-222, 2004.

[9] S. Kim and M. Park, "Effectiveness of person-centered care on people with dementia: a systematic review and meta-analysis," Clin Interv Aging, vol. 12, pp. 381-397, 2017.

[10] L.E. Legere et al., "Non-pharmacological approaches for behavioural and psychological symptoms of dementia in older adults: A systematic review of reviews," J Clin Nurs, Aug 9 [Epub ahead of print], 2017.

[11] G. Livingston et al., "Non-pharmacological interventions for agitation in dementia: systematic review of randomised controlled trials," Br J Psychiatry, vol. 205, no. 6, pp. 436-42, 2014.

[12] I. Abraha et al., "Systematic review of systematic reviews of non-pharmacological interventions to treat behavioural disturbances in older patients with dementia. The SENATOR-OnTop series," BMJ Open, vol. 7, no. 3, pp. 1-31, 2017.

[13] P. Van Bogaert et al., "SolCos model-based individual reminiscence for older adults with mild to moderate dementia in nursing homes: a randomized controlled intervention study," Journal of Psychiatric and Mental Health Nursing, 2016. 23(9-10): pp. 568-575.

[14] R. Woods et al., "REMCARE: reminiscence groups for people with dementia and their family caregivers - effectiveness and cost-effectiveness pragmatic multicentre randomised trial," Health Technol Assess, vol. 16, no. 48, pp. v-xv, 1-116, 2012.

[15] C. Gudex et al., "Consequences from use of reminiscence--a randomised intervention study in ten Danish nursing homes," BMC Geriatr, vol.. 10: pp. 33, 2010.

[16] B. Woods et al., "Reminiscence therapy for dementia," Cochrane Database Syst Rev, vol. 2, 2005.

[17] Y. Zhang et al., "Does music therapy enhance behavioral and cognitive function in elderly dementia patients? A systematic review and meta-analysis," Ageing Res Rev, vol. 35, pp. 1-11, 2017.

[18] M. Hsu et al., "Individual music therapy for managing neuropsychiatric symptoms for people with dementia and their carers: a cluster randomised controlled feasibility study," BMC Geriatr, vol. 15, pp. 1-19,, 2015.

[19] Y. Lin et al., "Effectiveness of group music intervention against agitated behavior in elderly persons with dementia," Int J Geriatr Psychiatry, vol. 26, no. 7, pp. 670-8, 2011.

[20] V. Grondahl et al., "The use of life stories and its influence on persons with dementia, their relatives and staff - a systematic mixed studies review," BMC Nurs, vol. 16: pp. 1-11, 2017.

[21] P. Subramaniam, B. Woods, and C. Whitaker, "Life review and life story books for people with mild to moderate dementia: a randomised controlled trial," Aging Ment Health, vol. 18, no. 3, pp. 363-75, 2014.

[22] A. van den Pol-Grevelink, J. S. Jukema, and C. H. Smits, "Person-centred care and job satisfaction of caregivers in nursing homes: a systematic review of the impact of different forms of person-centred care on various dimensions of job satisfaction," Int J Geriatr Psychiatry, vol. 27, no. 3, pp. 219-29, 2012.

[23] J. McKeown et al., "The use of life story work with people with dementia to enhance person-centred care," Int J Older People Nurs, vol. 5, no. 2, pp. 148-58, 2010.

[24] U. Kellett et al., "Life stories and biography: a means of connecting family and staff to people with dementia," J Clin Nurs, vol. 19, no. 11-12, pp. 1707-15, 2010.

[25] L. Chenoweth et al., "PerCEN trial participant perspectives on the implementation and outcomes of person-centered dementia care and environments," Int Psychogeriatr, vol. 27, no. 12, pp. 2045-57, 2015.

[26] J. Shore and S. Warden, The Art of Agile Development, Sebastopol, 2008.

[27] L. Madeyski and M. Kawalerowicz, Continuous Test-Driven Development — A Novel Agile Software Development Practice and Supporting Tool, in 8th International Conference on Evaluation of Novel Approaches to Software Engineering (ENASE), Angers. pp. 260-267, 2013.

Proposal of Field Oriented Event Messaging System

Yoshitoshi Murata, Rintaro Takahashi, Tomoki Yamato
Faculty of Software and Information Science
Iwate Prefectural University
Takizawa, Japan
e-mail: y-murata@iwate-pu.ac.jp, {g031n100, g031n161}@s.iwate-pu.ac.jp

Shohei Yoshida, Masahiko Okamura
Graduate School of Software and Information Science
Iwate Prefectural University Graduate School
Takizawa, Japan
e-mail: {g231n034, g231p005}@s.iwate-pu.ac.jp

***Abstract*— Several kinds of communication systems have been provided to hospitals in recent years. In these network systems, connected terminal equipment is managed and operated on a central console. This means that an operator is needed to do the job. To address this problem, we have proposed and developed a novel event messaging system for hospitals in which a sensor device paired with a mobile terminal reads a Quick Response Code (QR-code) on a display of sensor devices at a patient's bedside. When the device detects a patient is having trouble, it sends event messages to the terminal. The system enables hospital workers to take over the handling of patient monitoring operations by reading the original QR-code on their own mobile terminal.**

Keywords-hospital; communication system; nurse call; patient monitoring system; QR-code.

I. INTRODUCTION

Several kinds of communication systems have been provided to hospitals in recent years. They are roughly classified into call systems and remote patient monitoring systems. The latter are divided into event monitoring and data measuring types, which are sometimes integrated as a comprehensive network system. In Japan for example, Carecom Inc. provides a patient-nurse hotline system [1]. In this system, several kinds of sensor devices are connected to the hotline. One such device is a mat sensor to detect patients leaving their bed [2]. The mat is beside the bed and if a patient steps on it, an alert is sent to a nurse station. Honeywell provides a tracking and localization system integrated with a patient communication system and a call system [3]. General Electric Company (GE) provides many kinds of patient monitoring equipment [4]. They are connected to a central computer server through an intranet in a hospital. This makes it possible for medical professionals to monitor measured data.

In these network systems, connected terminal equipment such as patient monitoring equipment that allows measured data to be accessed is managed and operated on a central console. This means that an operator is needed to do the job, which is not always the case in emergency situations.

To address this problem, we propose a novel event messaging system in which a monitoring terminal by the patient's bedside is connected to a wireless relay unit such as a smartphone or PC. The wireless relay unit is connected to a remote monitoring terminal and a QR-code is read on the latter. This operation can be done at the bedside of a patient. We assume that this operation would be done by a medical worker such as a nurse who has a remote monitoring terminal. When the device detects a patient is having trouble, it sends a message to a remote monitoring terminal. A medical worker who receives the message would go to the monitored patient. We developed a wireless relay unit and a remote monitoring terminal using an Android smartphone. This makes it possible to hand the patient monitoring operation to another worker merely by having the latter read the QR-code on the preceding smartphone. This operation is the same as reading the original QR-code.

We developed two patient monitoring systems that use this messaging system. One monitors cases when intravenous feeding devices are removed from a patient and the other monitors cases when the patient leaves the bed.

After introducing related work, we will describe the concepts of this system in Section III. Practical monitoring systems are introduced in Section IV. Conclusions and future work are described in Section V.

II. RELATED WORKS

Remote monitoring systems for patients have been put to use ever since communication systems for sending messages from monitoring equipment to hospital personnel first started to be used.

One example is the "Risho Catch" system Paramount Bed Inc. has developed [5]. "Risho" means getting out of a bed in Japanese. This system detects when a patient sits up in bed, sits on the side of the bed, or leaves the bed and walks around the room. When this happens, it sends a message to a nurse station through a patient-nurse hotline system. The system structure is shown in Fig. 1.

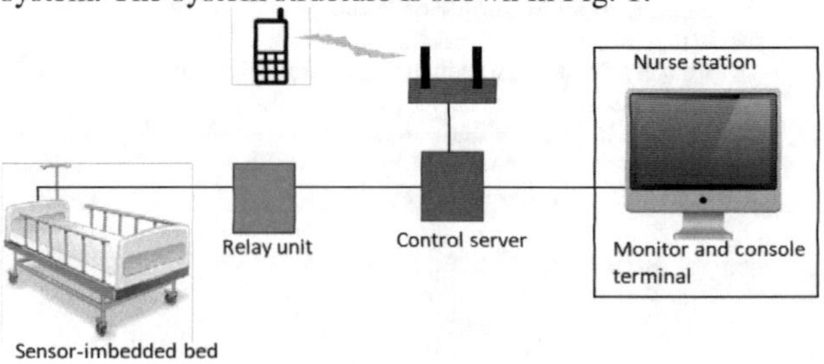

Figure 1. System and network structure of "Risho Catch."

A load sensor unit in the bed is connected to the patient-nurse hotline system through a relay unit located at the patient's bedside. The nurse station has a monitor and console terminal through which messages are sent to mobile devices.

Balaguera et al. evaluated decreasing the number of falls from bed using the SensableCare System [6]. Its architecture is shown in Fig. 2. The sensor pad in the system sends data through a cable to the control box located at the patient's bedside. The control box wirelessly transmits this data to Bluetooth routers located throughout the ward. This information then travels through the hospital network to the dashboard and docking server where the data is analyzed. When an alert is sent to the nurse via an application on his/her mobile phone, it is wirelessly transmitted through the hospital's Wi-Fi network. The patient's condition is monitored on the dashboard terminal.

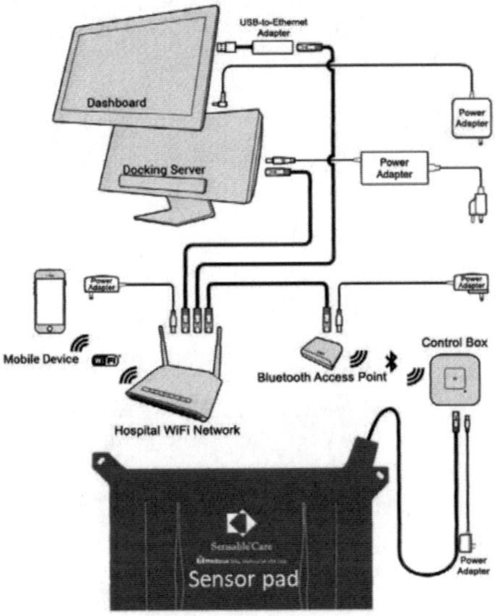

Figure 2. Architecture of the SensableCare System.

In both systems, the console terminal in the nurse station must be connected to the sensor unit, which the terminal must operate. This makes it hard to change monitoring terminals to other patients.

In the early 2000s, there were several remote patient monitoring systems that used the General Packet Radio Service/Wireless Local Area Network (GPRS/WLAN) as the wireless network and the Personal Digital Assistance (PDA) as the mobile device [7]-[10]. In these systems, sensor devices were connected to a monitor terminal or server through a wireless network. Since they were experimental systems, they had no fixed destination address. In recent years, several companies have been providing not only patient remote monitoring devices but also cloud services. GE provides a "GE Health Cloud" system along with many kinds of sensor devices and monitor devices [11]. The cloud manages the connecting of sensor devices to the hospital network and operates them on the console terminal. This scheme maintains a high security level but lacks flexibility. It makes it difficult for medical workers to install and pair sensor devices at the patient's bedside.

III. EVENT MESSAGING SYSTEM

In this paper, we describe the novel messaging system. We designed the system so that:

(1) Medical personnel could install sensor devices.
(2) They could pair the sensor devices with their own mobile terminals at the patient's bedside.
(3) Pairing situations could be monitored from a console terminal.
(4) Mobile terminals belonging to other organizations would be excluded.

Therefore, no operations are performed with the console terminal and monitoring sensors can be easily installed at the patient's bedside. The system configuration is shown in Fig. 3. The system consists of two programs; one is for the wireless relay unit and the other is for the mobile terminal. They collaborate with the Google Firebase Cloud Service (GFB) [12]. The GFB has many functions. This time, we use an authentication functions to exclude non-registration terminals, real time database to manage and monitor a status of pairing, and push messaging function to send notifications.

When a mobile terminal establishes pairing with a wireless relay unit, the ID of the wireless unit (WR-ID) must be entered. We used the QR-code for a worker to enter a WR-ID easily. A WR-ID is inputted merely by reading a QR-code on a display of a wireless relay unit. This system using a QR-code is useful for handing monitoring work to another worker. The handing operation is done by a worker reading a QR-code on the preceding mobile terminal. This QR-code is created from the WR-ID read from the wireless relay unit. In the case shown in Fig. 3, the first message "Mr. P has woken up." was sent to Worker A, and the second message "Mr. P has fallen." was sent to Worker B.

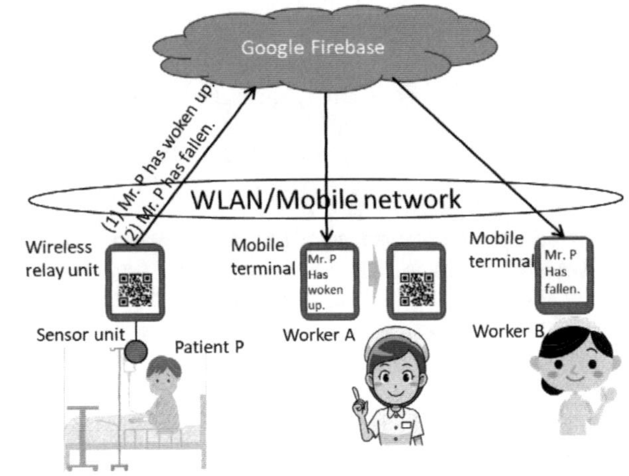

Figure 3. Configuration of the event messaging system.

The sequence to establish pairing between a wireless relay unit and a mobile terminal is shown in Fig. 4. After making a project to develop a messaging program, the GFB sends the Project code and Software Development Kit (Google-service. jason) to a developing PC (D. PC). These are installed to the wireless relay unit program and mobile terminal program. The wireless relay unit program creates a QR-code from the wireless relay ID (WR-ID). This is named by a developer or a user and is unique within an organization such as a hospital. When the mobile terminal program first accesses the GFB, a terminal ID (T-ID) is sent from the GFB. The mobile terminal program reads a QR-code on a wireless relay unit and sends the MT-ID and WR-ID to the GFB. The GFB makes a pairing between a wireless relay unit and a mobile terminal based on the Key. When a sensor device detects a change of state, the wireless relay unit of the sensor device sends a message to the GFB. The GFB then pushes the message to a paired mobile terminal.

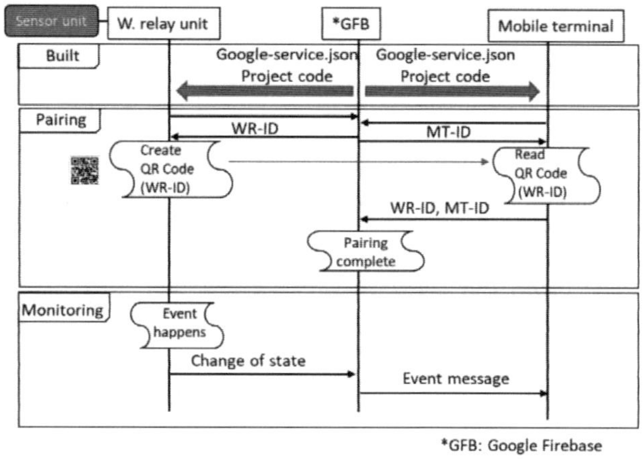

Figure 4. Sequence flow of pairing and monitoring.

IV. MONITORING APPLICATION

We developed two monitoring applications that adopt the proposed event messaging system. One is a intravenous drip monitoring application; the other is a fall monitoring application. We will describe their details in this section.

A. *Intravenous drip monitoring*

Some cognitive impairment patients sometimes remove an intravenous drip set by themselves. One existing intravenous drip monitoring system uses a switch type sensor that fastens a drip tube to detect when a drip set is removed as shown in Fig. 5 [14]. In case of a "Tenteki call", if a patient removes the switch type sensor together with the drip set, the sensor cannot detect the removal.

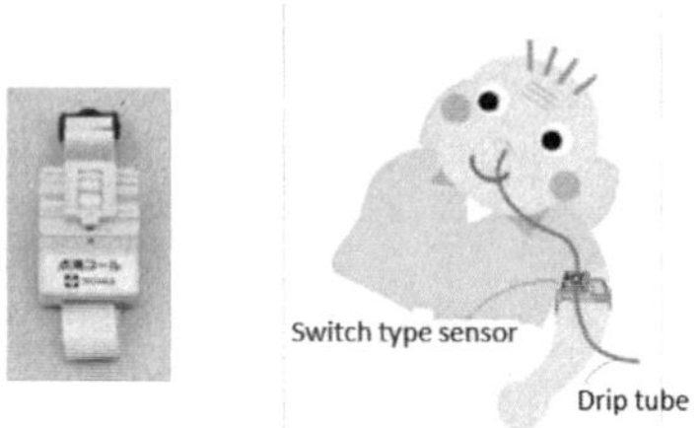

(a) Switch type sensor. (b) Example of use.

Figure 5. "Tenteki" call (Japanese), Technos Japan Co., Ltd.

We use a magnetic patch and a wireless magnetic sensor to remove a drip set, as shown in Fig. 6. The magnet is fastened to the body with an adhesive film in a place such as an arm. An intravenous drip tube is also fastened to the wireless magnetic sensor and the sensor is fastened to a magnetic patch with a medical fixing film. When a patient removes an intravenous drip set, the sensor is also removed from the body part. However, since the magnetic patch is fastened to the body part with an adhesive film, it must remain on the body part.

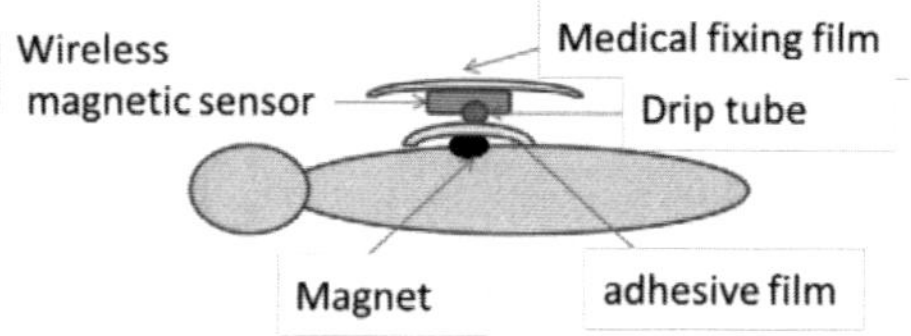

Figure 6. Structure

We developed the prototype system shown in Fig. 7. We used the STEVAL-WESU1 [15] developed by STMicroelectronics as the wireless magnetic sensor (see Fig. 7 (a)), and an Android smartphone as the wireless relay unit and mobile terminal. The muscle stiffness obtained by equipment manufactured by PIP Co., Ltd. was used as the magnetic patch [16] (see Fig. 7 (b)). In this prototype system, the program for detecting removal is integrated with the wireless relay program.

While a magnetic sensor is on a magnetic patch (see Fig. 7 (c)), the measured magnetic strength is bigger than the decision level (see Fig. 7 (e)). When the magnetic sensor is removed (see Fig. 7 (d)), the measured magnetic strength is less than the decision level; the detecting program has determined that an intravenous drip set has been removed (see Fig. 7 (f)). The wireless relay program sends a message "Yoshida's tube has been removed." That message is displayed on the mobile terminal (see Fig. 7 (g)).

B. *Fall monitoring*

Elderly people, especially cognitive impairment patients, have an increased risk of falling and consequently injuring themselves. They need to be prevented from falling to maintain their health because injuries from falling are a major reason for them to prolong their staying in a hospital.

(a) Magnetic sensor and attached tube. (b) Magnetic patch on an arm.

(c) Magnetic sensor on a magnetic patch. (d) A removed drip set.

(e) W. R. display :monitoring (f) W. R. display : detecting tube removing (g) M. T. display : receiving a message

Figure 7. Drip monitoring system prototype.

Therefore, many kinds of fall prevention systems have been developed. Most of them are classified into three schemes. The first type uses a mat type sensor like the systems described in Section II, the second one uses load sensors that are mounted in the legs of a bed, and the third one uses a camera. We developed a fall prevention system in which MS-KINECT was used. This is one of the third types. M. J. Rantz developed a fall detection system that uses MS-KINECT [17]. A medical worker monitors and judges whether a patient falls through the depth image of a patient on a monitor display.

On the other hand, our developed system detects whether a patient in a bed wakes up, sits up, stands up, or falls on the floor with a skeleton image of the patient. The detecting algorithms are as follows;

(1) Waking up: detecting that the head's height position is higher than the judging height 1.
(2) Sitting up: detecting both shoulders and a spine base angle of 25 or more degrees.
(3) Standing up: detecting the head, both shoulders and both hips, and a head is higher than the judging height 2.
(4) Falling down: detecting that the head's height position is lower than the judging height 3.

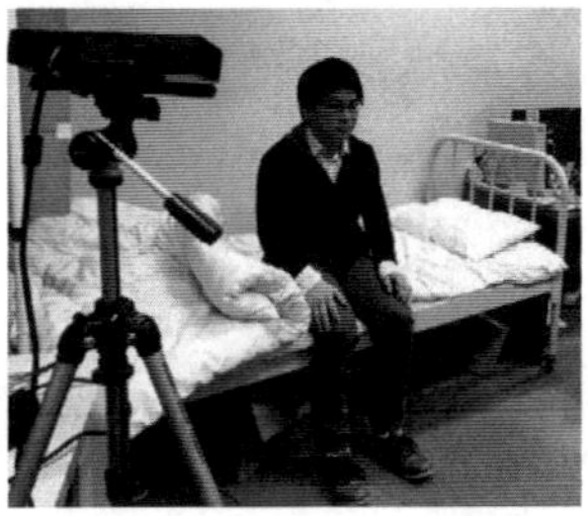

Figure 8. Experimental image.

We experimentally tested whether the developed system can detect the four conditions given above. The MS KINECT was positioned diagonally in front of the bed so that the front of the patient could be observed as shown in Fig. 8. Experimental results are shown in Fig. 9. Since MS-KINECT works on a PC, the sensor device program that detects patient conditions with MS-KINECT and the wireless relay unit program are combined on a PC. Monitoring images of participants as patients and the QR-code for pairing are shown on the PC display.

(1) Waking up (2) Sitting up

(3) Standing up (4) Falling down

(5) Screenshot of a mobile terminal that detects a "standing up" message

Figure 9. Experimental results for patient monitoring.

The developed system can detect four conditions for a patient. With it, a mobile terminal receives a “standing up” message sent from a wireless relay unit. However, the system sometimes makes mistakes in detecting conditions. Accordingly, we plan to improve our algorithms so that detection accuracy will be increased. We should also mention that a patient using this experimental system is presented with color images. We plan to change these color images to depth images to maintain a patient’s privacy.

V. CONCLUSION AND FUTURE WORK

We have developed a novel event messaging system in which a sensor device is paired with a mobile terminal by reading a QR-code on a display of sensor devices. Therefore, it is possible to pair them at the bedside of a patient. When the sensor devices detect a patient is having trouble, event messages are sent to a mobile terminal. We also developed a drip monitoring system and a fall monitoring system for use with the proposed event messaging system. The systems have not yet been used in hospitals. Accordingly, as a subject for future work, we plan to improve them and have them evaluated in hospitals.

REFERENCES

[1] Nurse Call BCAC, https://www.carecom.jp/global/solutions/nurse-call-bcac/ [retrieved: January, 2018].

[2] Mat sensor, Carecom, https://www.carecom.jp/global/solutions /option/ [retrieved: January, 2018].

[3] Clino Guard, Ackermann, Honeywell, https://www.ackermann-clino.com/en/products/tracking-localization/ [retrieved: January, 2018] .

[4] Patient Monitors, GE, http://www3.gehealthcare.com/en/products/categories/patient_monitoring/patient_monitors [retrieved: January, 2018].

[5] Risho catch, Paramount Bed, http://www.paramount.co.jp/learn/inversion/leaving [in Japanese, retrieved: January, 2018].

[6] H. U. Balaguera et al., “Using a Medical Intranet of Things System to Prevent Bed Falls in an Acute Care Hospital: A Pilot Study,” Journal of Medical Internet Research, https://www.jmir.org/2017/5/e150/, [retrieved: January, 2018].

[7] A. V. Halteren et al., “Mobile Patient Monitoring: The MobiHealth System,” The Journal on Information Technology in Healthcare 2004, vol. 2(5), pp.365–373, 2004.

[8] Y. Lin et al., “A Wireless PDA-Based Physiological Monitoring System for Patient Transport,” IEEE Transactions on Information Technology in Biomedicine, Vol. 8, No. 4, pp.439-447, 2004.

[9] T. Gao et al., “Vital Signs Monitoring and Patient Tracking Over a Wireless Network,” Johns Hopkins APL Technical Digest, Vol. 27, No. 1, pp. 66-74, 2006.

[10] P. Várady, Z. Benyó, and B. Benyó, “An Open Architecture Patient Monitoring System Using Standard Technologies,” IEEE Transactions on Information Technology in Biomedicine, Vol. 6, No. 1, pp. 95-98, 2002.

[11] Patient monitoring, GE Healthcare, http://www3.gehealthcare.com/en/products/categories/patient_monitoring, [retrieved: January, 2018].

[12] Google Firebase Cloud Messaging, https://firebase.google.com/docs/cloud-messaging/?hl=en, [retrieved: January, 2018].

[13] QR-code, http://www.qrcode.com/en/, [retrieved: January, 2018].

[14] Tenteki call (Japanese), Technos Japan Co., Ltd., http://www.technosjapan.jp/product/infusion/2012/0121195451.html [in Japanese, retrieved: January, 2018].

[15] STEVAL-WESU1, STMicroelectronics, http://www.st.com/en/evaluation-tools/steval-wesu1.html [retrieved: January, 2018].

[16] Muscle Stiffness, http://www.pip-club.com/english/muscle.html, [retrieved: January, 2018].

[17] M. J. Rantz et al, “Automated Fall Detection With Quality Improvement “Rewind” to Reduce Falls in Hospital Rooms,” NIH Public Access, Journal of Gerontological Nursing, 40(1), pp. 13–17, 2014.

Automated Assessment of Nonverbal Behavior of the Patient during Conversation with the Healthcare Worker Using a Remote Camera

Takashi Imabuchi, Oky Dicky Ardiansyah Prima, Hisayoshi Ito
Faculty of Software and Information Science
Iwate Prefectural University
Takizawa, Iwate, Japan
e-mail: g236o001@s.iwate-pu.ac.jp, {prima, hito}@iwate-pu.ac.jp

***Abstract*—The importance of nonverbal behavior between the healthcare worker and the patient has led many studies to focus on quantitatively evaluating this behavior. Those existing studies utilize the Kinect 3D sensor to visualize the nonverbal behaviors for a range of healthcare applications. In this study, we propose a framework to automatically assess the nonverbal behavior of the patient during conversation with the healthcare worker. Instead of using the Kinect, we detect the skeleton information of the targeted body from a single camera using an advanced computer vision approach. The proposed framework collects data consisting of facial expression, eye movement, and head nodding and analyzes this data to assess the quality of the nonverbal communication.**

Keywords-Nonverbal communication; Eye tracking; Facial expression; Pose estimation.

I. INTRODUCTION

In the health sector, good communication between healthcare workers and patients is important to improve the quality of care and to promote patient-centered healthcare. These communications are effective to improve the patient satisfaction [1]. Assessing the quality of communication in a healthcare setting is a difficult task in today's hospitals. The assessment includes both verbal and nonverbal communications [2]. While the verbal communication directly conveys the patient's needs, the nonverbal communication represents communicative acts which may be even more important than the matter under verbal discussion. The nonverbal communication consists of a variety of non-words information such as gestures, physical features and paralanguages. The nonverbal interaction in the healthcare sector may represent as much as 65 percent of the hidden thoughts and emotions of the patient [3]. To establish a good relationship of trust with a patient, the healthcare worker is required to pay close attention to this interaction. This skill would be essential for all healthcare workers.

There are basically five types of nonverbal behaviors (body language) related to movement of the body [4]: emblems, regulators, illustrators, affective display and adaptors. During a communication with patients, skilled healthcare workers (health professionals) use four elements of the body language: body posture, eye contact, facial expression, and gesture. Hence, the quality of communication can be observed from the appearances of the visual components of their faces in accordance with the movements of other parts of their bodies.

Information and Communications Technologies (ICTs) play an important role to automatically assess the nonverbal communication. Depth camera sensing enables the observation of human pose in three dimensions. The Kinect 3D sensor shows adequate performance for a range of healthcare imaging applications [5]. It provides not only the changes of the body posture, but also the pose (skeleton) information of the targeted body, which can be used to estimate gestures. Facial image processing enables the detection of head movements and the subtle changes of facial expression. Using Internet-of-Things (IoT) [6], these tasks can be processed with a low investment cost.

The major challenge of implementing camera sensing into the healthcare sector is determining where to place the camera so that it can capture both the healthcare worker and the patient at the same time, with adequate image resolution. Since the camera's existence must not disturb the process of care, it should be placed around corners or on the ceiling in the hospital room. However, placing the camera far away from the target will decrease its ability to reveal the detail structures of the target's body. An attempt using a combination of multiple cameras, depth sensing, and fiducial markers has been conducted to measure hand movements of the healthcare worker with respect to the patient's bed [7]. While this approach enables to estimate the potential hand movements at the bedside, the overall system is complicated for a practical use in a hospital.

Recent computer vision applications enable the detection of 2D human poses from a single image [8]. Furthermore, the 3D human pose can be estimated by using human pose libraries taken from motion capture devices as a reference [9]. Unlike the Kinect which needs a proper distance setting to the targets, these approaches are more flexible. The skeleton information can be derived for the targeted bodies located more than 5m away from the camera. Based on this information, the detection of the head and the visual components of the face can be analyzed easier without requiring face detections as in the conventional image processing [10].

In this study, we propose a framework to automatically assess the nonverbal behavior of the patient during conversation with the healthcare worker from a single camera. The proposed framework starts analysis at the time it detects a skeleton other than that of the patient. Using the part of the skeleton that consists of the bones of the head (axial skeleton), the face area of the patient is determined and up-sampled to

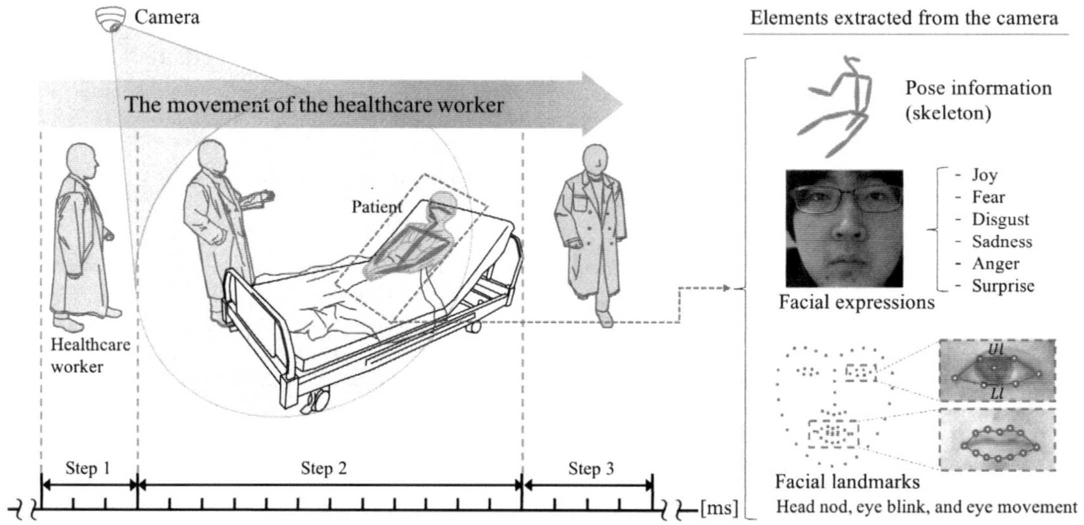

Figure 1. The proposed framework to assess nonverbal behaviors from a single camera.

observe the facial information in detail. Information consisting of facial expression, eye movement, and head nodding is statistically analyzed against the pre-calculated learning data to show how well the nonverbal communication is being constructed. For further application, we also discuss other possibilities on the uses of skeleton information to improve the quality of healthcare service.

This paper is organized as follows. Section II describes related works in healthcare sector. Section III introduces our approach to assess nonverbal behavior using a remote camera. Section IV describes our experimental setting and its results. Finally, Section V presents our conclusions and future works.

II. RELATED WORK

Traditional methods to measure nonverbal behavior rely on manual coding system. This measurement involves the duration, the response latency, and the inter-response time during a behavior [11]. Since there are a lot of ambiguities on judging a particular behavior, many observers tend to perform in an inconsistent manner, thus degrading the quality of the measurements. Some studies have developed automated detection of the nonverbal behavior based on velocities changes derived from gyroscopes [12], video-based motion analyses, and a multi-modal sensor consisting of red, green and blue (RGB), depth, and audio data. For a more complex and dynamic behavior, high-level sensing is required to assess [13].

In the health sector, many studies have been conducted to assess the nonverbal behavior between the healthcare worker and the patient. These assessments covered conversational agents, securities, healthcare settings and medical acts in the hospital room. The Kinect 3D sensor has been used to extract and analyze head pose and hand gestures of the healthcare worker. The typical hand gestures can be classified using machine learning algorithm [14]. The depth sensing in the Kinect is also shown to be adequate to monitor the respiratory rate of the patient [15].

Although the Kinect 3D sensor provides pose information to measure the high-level behavior, it has drawbacks for a practical use in a hospital. The measurement of nonverbal behavior in the hospital setting requires a sensor to capture a wider area where the healthcare worker and the patient are located. Moreover, operating the Kinect and its processing computer will incur a high investment cost.

The advances brought about by machine learning and Artificial Intelligence (AI) have contributed to solve the problem on localizing anatomical key points to find body parts inside an image. The "OpenPose" library enables realtime multi-person 2D pose estimation from a single camera [8]. This library applies a bottom-up approach by encoding the location and rotation of limbs over the image domain to allow a greedy parse to connect the detected body parts. Thus, the library can detect human pose not only from the front but also from the back. Martinez et al. [16] used deep neural networks to map between 2D and 3D poses. The standard protocol of Human3.6M was used to normalize the pose data [17]. Human pose estimation from a single camera has become a potential alternative to the Kinect.

III. AUTOMATED ASSESSMENT OF NONVERBAL BEHAVIOR

A. *Nonverbal Behavior Measurement*

Here, we assume a situation in a hospital where the nonverbal communication between the healthcare worker and the patient mostly occurs. Figure 1 shows three steps for our proposed method to assess the nonverbal behavior. A single camera is located facing the patient's bed. For each step, we use the "OpenPose" library to detect human pose found in the camera image. A pre-defined region of interest (ROI) is applied to the image to assign the location of the bed.

Step 1: Detecting the presence of the healthcare worker

When the healthcare worker is heading towards the patient's bed, his posture will be detected by the time he stands in front of the bed. "OpenPose" detects the existence of human pose inside ROI. It enables to distinguish the basic body postures: standing, sitting (reclining), and laying. The patient is recognized to have laying or reclining postures while the health worker to have a standing posture. Once the healthcare worker is detected when he enters the ROI, he will be tracked until he leaves the area.

Step 2: Extracting nonverbal behavior during conversation

Nonverbal behaviors of the patient during speaking and listening are extracted separately. The periods of speaking and listening are defined by utterances, occurrences of changes in shapes of the patient's mouth detected from the camera image [18]. To calculate the facial shape, 68 facial landmarks are estimated using "Dlib" library [10]. The extracted nonverbal behaviors include head nod, eye blink, eye movement, and facial expressions. While the calculations of head nod, eye blink, and eye movement make use of facial landmarks, the facial expression is estimated using "Affectiva" library, a state-of-the-art emotion recognition using Deep Learning [6].

Step 3: Detecting the end of conversation

The observation will finish when the healthcare worker leaves the room as detected by the absence of his posture in the camera image.

The details of methods for detecting utterances, head nod, eye blink, eye movement, and facial expressions are as follows.

1) Utterance: Utterance is measured by detecting the relative location changes of the boundary of the mouth represented by 12 landmarks. These changes are observed as the standard deviation of those landmarks within a duration. Utterance is quantized by

$$Utterance = \frac{1}{n}\sum_{i=1}^{n}\sqrt{\frac{1}{m}\sum_{j=1}^{m}\left(L_{nj(x,y)} - \overline{L_n}\right)^2}, \quad (1)$$

where, n is the number of image frames, m is the number of landmarks, and L_{nj} is j-th landmark point (x, y) at n-th frame.

2) Head nod: Head nod is calculated by detecting the changes of head-pose in pitch angle. This calculation is done by fitting six 3D anthropometric points to the associated facial landmarks. The changes of pitch angles are calculated by

$$Head\,nod = \sqrt{\frac{1}{n}\sum_{i=1}^{n}(\varphi_n - \bar{\varphi})^2}, \quad (2)$$

where, n is the number of image frames and φ_n is the *pitch* angle at n-th frame.

3) Eye blink: The eye blink is considered as the degree of eye openness which can be measured from the relative ordinate changes of the midpoints of upper and lower eyelids represented by previously calculated landmarks by

$$Eye\ openness = \sqrt{(Ul_y - Ll_y)^2}, \quad (3)$$

where, UL_y and Ul_y are ordinates of midpoints of upper and lower eyelids, respectively.

4) Eye movement: We measure the eye movement by tracking the iris inside the region of eye in the image. The region of eye is extracted from the area of image surrounded by eye landmarks. The tracking method is based on eye tracking from visible-spectrum camera [19]. This method calculates the gaze direction with head pose compensation. Let (x, y) the coordinate of the iris center, horizontal and vertical eye movements (E_{dx}, E_{dy}) in n-number of frames are calculated by

$$E_{dx} = \sqrt{\frac{1}{n}\sum_{i=1}^{n}(ex_i - \bar{e})^2}\ , \quad E_{dy} = \sqrt{\frac{1}{n}\sum_{i=1}^{n}(ex_i - \bar{e})^2}. \quad (4)$$

5) Facial expression: The "Affectiva" library estimates six fundamental expressions: joy, fear, disgust, sadness, anger, and surprise from the patient face in the image. The occurrence of each expression is represented as a probabilistic value. Among these expressions, we use only the accumulated occurrence of "joy" for our purpose, as described later on this paper.

B. Assessment of nonverbal behaviors

Our strategy to assess nonverbal behaviors is to create two scenarios of communication scenes between the healthcare worker and the patient. The first scenario is "high trust" where a good quality relationship between the healthcare worker and the patient has been established. Contrary, the second is "low trust" scenario. While the "high trust" will lead to greater emotional stability that facilitates acceptance and openness of expression, "low trust" will result in less accurate communication. We believe that these differences can be revealed from the nonverbal behaviors extracted in this study.

Figure 2 shows our preliminary results of nonverbal behaviors of the two scenarios. Both manual and automatic coding of utterances are provided to show our utterance detection is effective to define the cognitive tasks: listening and speaking. Nonverbal behaviors for each scenario are assessed as follows.

1) Repetition of head nod, eye blink, and eye movement: Here, to distinguish communicative from noncommunicative types head nod, eye blink, and eye movement, we count the repetition for those behaviors in a duration of time and in each cognitive task: listening or speaking. The number of repetitions is computed by counting the number of values

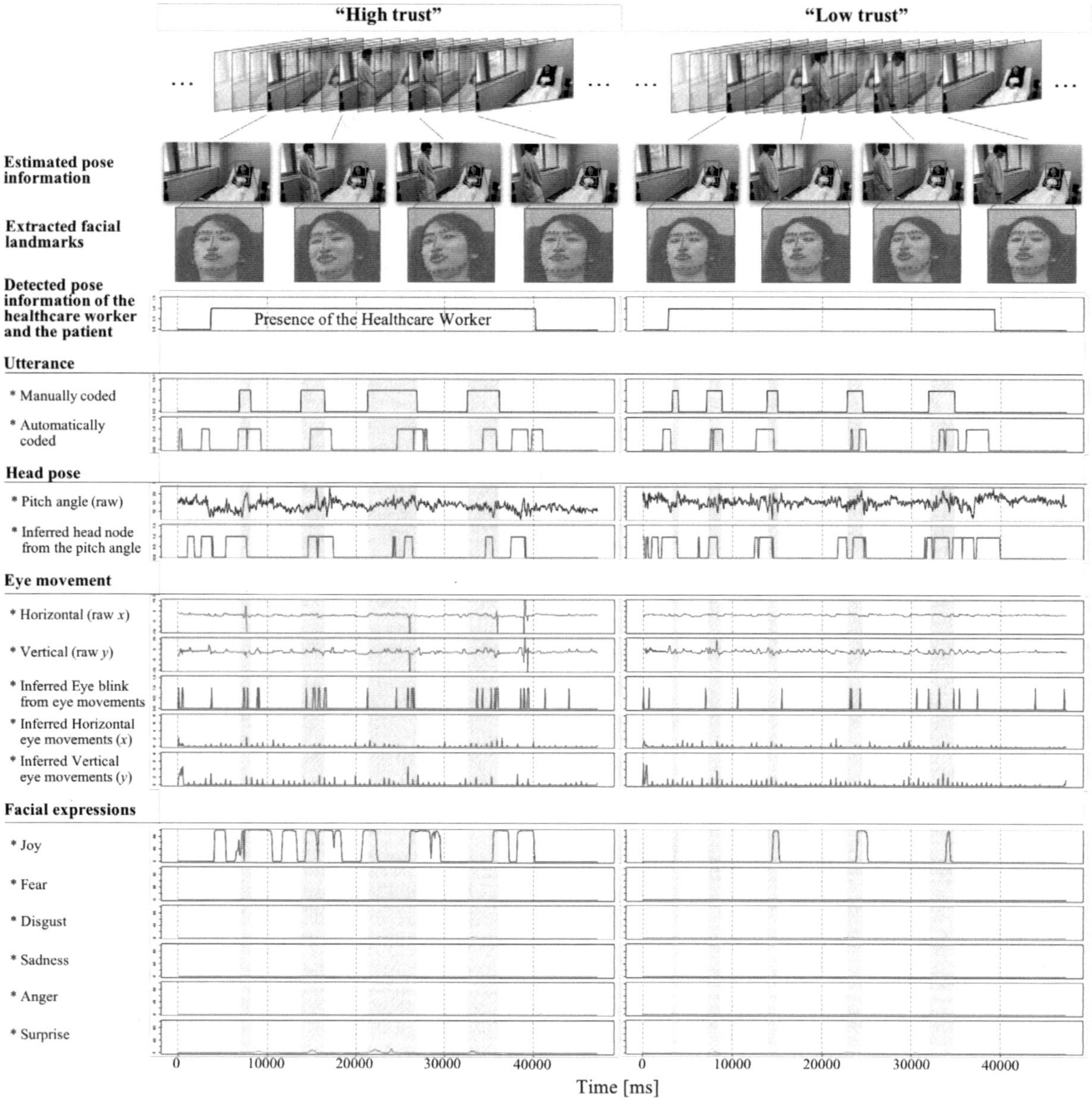

Figure 2. Preliminary results of nonverbal behaviors of the two scenarios.

above a given threshold in (1), (2), (3), and (4). Different patterns for head nod, eye blink, and eye movement during listening and speaking can be observed, as shown in the filled and unfilled area in Figure 2.

2) Accumulated occurrences of facial expressions: Since facial expression will have a range of amplitude, occurrence and duration, we calculate the probability accumulation for each facial expression in each cognitive task. Figure 2 shows different patterns of six fundamental expressions during speaking and listening. Since the fluctuations of "joy" show high responses, for further analyzes, only "joy" will be calculated to represent the facial expression changes in this study.

IV. EXPERIMENTS AND RESULTS

Experiments were conducted by an experimenter and three participants as patients to communicate in "high and low trust" scenarios. Both scenarios were conducted for a duration of 40 seconds. The experimenter talked with the participant, in the way to cause the participants to alter their behavior to match each scenario. All scenes were recorded using a 60fps single camera with 1280×720 pixels resolution.

TABLE I. FREQUENCIES OF HEAD NOD, EYE BLINK, AND EYE MOVEMENT IN TWO SCENARIOS

Nonverbal Behaviors	Speaking [Hz]						Listening [Hz]					
	Subject #1		Subject #2		Subject #3		Subject #1		Subject #2		Subject #3	
	LT	HT	LT	HT	LT	HT	LT	HT	LT	HT	LT	HT
Head nod	0.4	0.2	0.0	0.1	0.6	0.5	0.2	0.4	0.1	0.3	0.3	0.1
Eye blink	0.5	0.6	0.2	0.5	0.6	1.1	0.7	0.4	0.4	0.3	0.2	0.1
Eye movement:												
-Horizontal (*x*)	1.1	1.0	0.6	1.0	1.0	1.0	0.9	0.4	0.3	0.7	0.9	0.8
-Vertical (*y*)	1.3	1.1	1.4	1.3	1.8	1.3	0.9	1.2	0.5	0.9	1.2	0.8

LT: "Low trust", HT: "High trust"

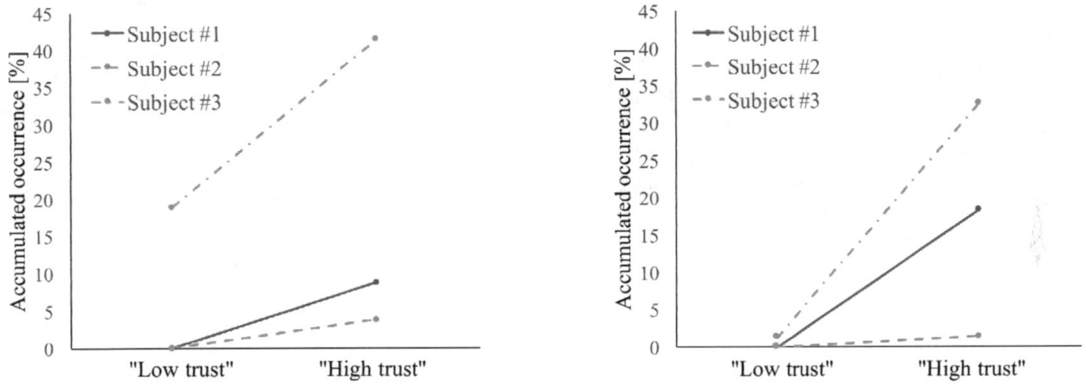

(a) Speaking (b) Listening

Figure 3. Accumulated occurrences of facial expression "joy" in two scenarios.

Table I shows frequencies of head nod, eye blink, and eye movement of the participants detected in two scenarios. Nodding frequency does not show significant patterns. Although patterns of nodding frequency are expected to change as the communication proceeds from "low trust" to the "high trust", we consider that the communication duration is too short to create emphatic responses of the participants. For further observation, it is necessary to break down the communication of each scenario into stages (initial stage, exploration stage, struggling stage, and closing stage). The frequency of eye blink shows a significant increase from "low trust" to the "high trust" during speaking. Conversely, it shows a significant decrease during listening. This result is consistent with the previous research where during tasks requiring higher cognitive load, subjects' blink rate tends to decrease [20]. The frequency of eye movement shows a significant increase in the vertical direction regardless of the scenarios. This behavior is highly affected by the experimental setting, where the participants' heads were in lower position than the experimenter. Therefore, when the participants tried to maintain eye contact with the experimenter, the vertical gaze movements were remarkable than the horizontal.

Figure 3 shows the accumulated occurrences of facial expression "joy" in two scenarios. The expression of "joy" increases from the "low trust" to the "high trust" scenarios in both speaking and listening behaviors. This result represents the feeling of being nervous or uncomfortable of the participant during communication in the "low trust" scenario.

V. CONCLUSION AND FUTURE WORK

We have demonstrated our framework to assess the nonverbal behavior of the patient from a single camera. With its ability to detect human pose information and facial landmarks, the presence of the healthcare worker can be detected to start analyzing the nonverbal behavior of the patient automatically. The proposed framework will also eliminate the need for a number of sensor devices such as gyroscopes and eye trackers, to analyze the nonverbal behavior.

The proposed framework can be extended to monitor the activity of the patients. For example, by using the detected pose information, the patient body movements such as sitting posture, posture of turning over in bed, and posture of getting down from the bed can be monitored and characterized for each patient. This information will lead to an improvement of

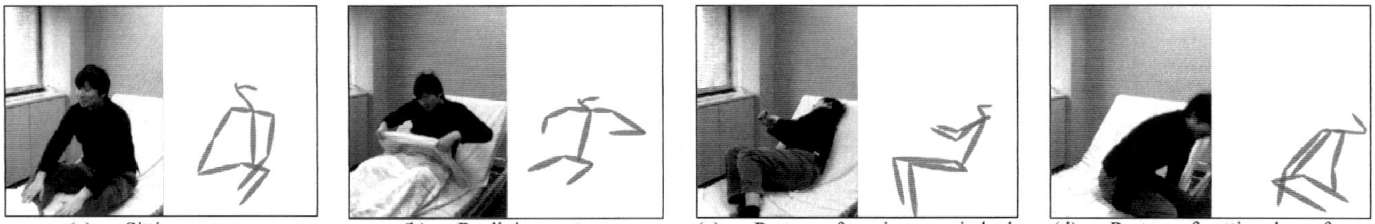

(a) Sitting posture (b) Reclining posture (c) Posture of turning over in bed (d) Posture of getting down from the bed

Figure 4. Using pose information to detect patient's body movements.

the healthcare service. Figure 4 shows four different postures of a user detected by our framework.

We will continue to conduct more experiments with various settings to confirm the robustness of our framework and to increase its functionality in order to improve the communication between the healthcare worker and the patient in particular and the healthcare service in general. By building a private cloud environment, data derived by our framework can be stored as big data to analyze possible efforts to increase patient satisfaction. Authorized persons can access the results by using cloud applications on hand-held devices. As a result, any issues related to the communication between the healthcare worker and the patient can be found at the early stage.

REFERENCES

[1] P. G. Northouse and L. L. Northouse, Health Communication: Strategies for Health Professionals, second ed., Northouse, 1992.

[2] L. S. Pettegrew, Some boundaries and assumptions in helthcare communication, Explorations in provider and patient interaction, Louisvillem KY:Humana, 1982.

[3] A. J. Davis, Listening and Responding : Cornerstone of Helping, Elsevier Health Sciences, 1984.

[4] P. Ekaman and W. V. Friesen, "The repertoire or nonverbal behavior: categories, rigins, usage, and coding," Semiotica, vol. 1, pp. 49–98, 1969.

[5] S. T. L. Pohlmann, E. F. Harkness, C. J. Taylor, and S. M. Astley, Evaluation of Kinect 3D Sensor for Healthcare Imaging, Journal of Medical and Biological Engineering, vol. 36, pp. 857–870, 2016.

[6] S. Thibaud, M. Daniel, and K. Rana, "Facial Action Unit Detection Using Active Learning and an Efficient Non-linear Kernel Approximation," 2015 IEEE International Conference on Computer Vision Workshop (ICCVW), 2015.

[7] J. Chen, J. F. Cremer, K. Zarei, A. M. Segre, and P. M. Polgreen, "Using Computer Vision and Depth Sensing to Measure Healthcare Worker-Patient Contacts and Personal Protective Equipment Adherence Within Hospital Rooms," *Open Forum Infect Dis*, vol. 3, no. 1, pp. 200–206, Feb. 2016.

[8] Z. Cao, T. Simon, S. E. Wei, and Y. Sheikh, "Realtime multi-person2D pose estimation using part affinity fields," Computer Vision andPattern Recognition, 2017.

[9] H. C. Chen and D. Ramanan, "3D Human Pose Estimation = 2D Pose Estimation + Matching," arXiv:1612.06524v2 [cs.CV], pp. 1–9, 2017.

[10] V. Kazemi and J. Sullivan, "One millisecond face alignment with an ensemble of regression trees," Proceedings of the IEEE Conference on Computer Vision and Pattern Recognition (CVPR 2014), pp. 1867–1874, 2014.

[11] J. O. Cooper, T. E. Heron, and W. L. Heward, Applied Behavior Analysis 2nd ed., Pearson, (2007).

[12] M. Inoue, T. Irino, N. Furuyama, R. Hanada, T. Ichinomiya, and H. Masaki, "Manual and accelerometer analysis of head nodding patterns in goal-oriented dialogues," Interaction Techniques and Environments - 14th International Conference, HCI International 2011, Orlando, FL, pp. 9–14, 2011.

[13] P. Turaga, R. Chellappa, V. S. Subrahmanian, and O. Udrea, "Machine Recognition of Human Activities: A Survey," IEEE Trans. Circuits Syst. Video Technol., vol. 18, no. 11, pp. 1473–1488, 2008.

[14] A. S. Won, J. N. Bailenson, S. C. Stathatos, and W. Dai, "Automatically Detected Nonverbal Behavior Predicts Creativity in Collaborating Dyads," Journal Nonverbal Behavior, vol. 38, no. 3, pp. 389–408, 2014.

[15] N. Burba, M. Bolas, and D. M. Krum, "Unobtrusive measurement of subtle nonverbal behaviors with the Microsoft Kinect," Proceeding VR '12 Proceedings of the 2012 IEEE Virtual Reality, pp. 1–4, 2012.

[16] J. Martinez, R. Hossain, J. Romero, and J.J. Little, "A simple yet effective baseline for 3d human pose estimation," arXiv:1705.03098 [cs.CV], pp. 1–10, 2017.

[17] C. Ionescu, D. Papava, V. Olaru, and C. Sminchisescu, "Human3.6M: Large Scale Datasets and Predictive Methods for 3D Human Sensing in Natural Environments," IEEE Transactions on Pattern Analysis and Machine Intelligence, vol. 36, no. 7, pp. 1325 1339, 2014.

[18] Y. Jaana, O. D. A. Prima, T. Imabuchi, H. Ito, and K. Hosogoe, "The development of automated behavior analysis Software," Proc. SPIE 9443, Sixth International Conference on Graphic and Image Processing (ICGIP), 2014.

[19] T. Imabuchi, O. D. A. Prima, and H. Ito, "Visible Spectrum Eye Tracking for Safety Driving Assistance," in Trends in Applied Knowledge-Based Systems and Data Science, vol. 9799, no. 4, pp. 428–434, 2016.

[20] L. Vincze and I. Poggi, Communicative Functions of Eye Closing Behaviours, Analysis of Verbal and Nonverbal Communication and Enactment. The Processing Issues Lecture Notes in Computer Science, vol. 6800, 2011.

Transition to Adult Care for Youth with Type 1 Diabetes Mellitus

Structured Support Aids for Adolescents and Young Patients in Morioka Area, Japan

Kazuma Takahashi
Department of Basic Nursing Science
School of Nursing, Graduate School of Nursing, Health Support Centre, Iwate Prefectural University
Takizawa, Japan
E-mail: t_kazuma@iwate-pu.ac.jp

Akio Takahashi
Department of Diabetes and Endocrinology
Morioka Children's Hospital
Morioka, Japan
E-mail: akiotakaha@m-codomo-hp.or.jp

Yasushi Ishigaki
Department of Diabetes and Metabolism
School of Medicine, Iwate Medical University
Morioka, Japan
E-mail: ishigaki@iwate-med.ac.jp

***Abstract*—In clinical practice for adolescents and young adults with childhood onset diseases, transitional medicine represents an important element. Type 1 Diabetes Mellitus (T1DM) is a disease that requires patients, their families, and health care professionals involved to face a wide range of problems, including early development of complications, decreased quality of life, socioeconomic problems, and developmental psychological problems. The transition to adult health care is often not smoothly carried out for young patients with this disease. In this paper, first, we discuss problems in this area, including differences between paediatric and adult clinical settings for diabetes care, insufficient preparation for transition, and dissatisfaction of patients and families during transition. Next, we propose a clinical trial, providing a seamless transitional process to the young patients with T1DM at a joint clinic operated by both pediatricians and physicians, trying to ascertain the changes in psychological burden on the patients, quality of life due to transition, self-care behaviors, and blood glucose control during the transition.**

Keywords-Type 1 Diabetes Mellitus; Transfer; Transition; Joint Clinic; Bridge Clinc.

I. Introduction

Progress in medicine over the last decades has given rise to increased survival of children with complicated paediatric diseases who reach adulthood [1]. The transfer of adolescents and young adults from paediatric to adult health care is a complex operation that may generate various difficulties for all participants. The actual operation is called "transition". This process requires young patients to get skills to increase their self-determination and become more responsible and prepared to face self-care, and for their families and health care professionals to be inclined to guide them [2].

Importance of transitional medicine has been recognized throughout the world, but discussion and consensus formation between paediatricians and physicians seem still inadequate. It is an urgent issue to prepare transition systems so that the adolescent and young adult patients can receive appropriate transitional process, considering changes of patients' status, progression of complications, and physical and personality maturity along with age. There are, however, the limited number of researches on the actual condition and problems of the transition of young patients with chronic diseases. Specialised medical facilities bridging transition are globally rare, as well. It is crucial to promote researches on this issue from the socioeconomic or psychological points of view, as well as from a clinical one.

In this article, first, we review and discuss problems in transition, focusing on Type 1 Diabetes Mellitus (T1DM), formerly referred to as juvenile diabetes mellitus, and then present a project to establish a transition system to cohere paediatric and adult health care for adolescent and young adults with T1DM around Morioka area in Iwate, Japan.

II. Type 1 Diabetes Mellitus

Type 1 Diabetes Mellitus arises because of β-cell destruction in the pancreas [3]. It is supposed that 415 million people across the world have diabetes and that T1DM accounts for approximately 7–12% of the subjects [4]. This disease develops most frequently during infancy and adolescence, in people under 30 years. Both genetic and environmental factors may play crucial roles, however, the exact pathogenesis of T1DM remains still unclear.

Beta cells produce a hormone, insulin, which regulates blood glucose levels. Since endogenous production of insulin is generally absent or in very small quantities in the patients with T1DM, lifelong treatment with insulin, either by multiple injections or insulin pump, and with frequent self-monitoring of blood glucose (SMBG) is required. Self-care is central to successful outcomes for individuals with T1DM and good diabetes management has been shown to minimise the risks of long-term and short-term complications [5].

Type 1 Diabetes Mellitus is a complex and demanding condition, which places a substantial behavioural and psychological burden on young people and their families. Diabetes management imposes considerable requirements, including nutrition guidelines, insulin regimens, glucose fluctuations, and SMBG, that are difficult for the patients to negotiate, specifically for children [6]. The transition into adolescence is often associated with poorer adherence to treatment, deteriorating metabolic control, and increased risk for psychological disorders [5]. In children with T1DM, parents are ultimately responsible for daily management of this disease, which can contribute to parental stress, distress, and diminished quality of life (QOL). Therefore, ensuring quality of life for the patients and their families, while maintaining glycemic control within targets, is an important challenge in diabetes treatment [7]. In addition, medical expenses also represent a large burden to both the patients and their families. Indeed, poor glycemic control in paeditric T1DM is associated with lower socioeconomic status [8].

III. EMERGING ADULTHOOD

The developmental period from ages 18–25 years has been termed ''emerging adulthood'' [9]. During this stage, many young adults leave their family homes, become financially independent, start working, and seek intimate relationships with partners [10]. Competing academic, economic, and social priorities often detract from engagement to chronic disease management. Even if young adults face these competing demands, they have not achieved all of the skills necessary to remain independent and accept these responsibilities on their own [9]. In addition, during this relatively healthy time in their lives young adults feel markedly invulnerable, and have a tendency to reject adult control.

The transition to emerging adulthood may give a tall order to those with T1DM. During this stage, on-going family involvement in diabetes management is necessary to reduce the risk of deterioration in glycemic control that often accompanies adolescence [11]. On the contrary, in addition to the new responsibilities and freedoms, those managing T1DM become more and more responsible for their self-care, with daily requirements including SMBG and insulin dose adjustments, and routine activities such as scheduling physician appointments, ensuring availability of sufficient stocks, and taking care of themselves when sick. It may be unrealistic to expect the person with diabetes in the early phase of emerging adulthood to make major changes in their diabetes management strategies, or even to transition to a new adult diabetes health care provider [12]. The demands of managing the complicated illness like T1DM must be integrated into the standardised changes in occupation, education, relationships, and living situations that accompany emerging adulthood [13].

IV. CULTURAL DIFFERENCE BETWEEN PAEDIATRIC AND ADULT DIABETES CARE

The management of T1DM in paediatric care differs in various ways from adult setting. In general, paediatricians are not accustomed to the needs of young adults, and physicians are trained with different viewpoints of diabetes care. Furthermore, the psychosocial environment and expectations differ markedly between two settings, and paediatric and adult providers have different ideas on transition [14]. Indeed, the adolescents and young adults experienced cultural difference during the transition. While children's care was generally characterized by close, supportive relationships with staff, lower anticipation of patient responsibility, and significant parental involvement, adult care was considered less supportive and less personal, with frequent changes in staff and decreasing involvement by parents [15]. This shift in culture was sometimes experienced as very precipitate [16]. Some felt out of place in adult services, many were less satisfied with adult care than children's care [17].

V. IMPACT OF TRANSITION ON GLYCEAMIC CONTROL, ATTENDANCE AT HEALTHCARE APPOINTMENTS, AND DIABETES-RELATED HOSPITALIZATIONS

A systematic review or meta-analysis by Sheehan et al. [18] tried to examine the impact of transition from child to adult health care on health outcomes and health behaviours for young people with child-onset T1DM.

In nine observational studies that assessed the impact of healthcare transition on glycaemic control either by examining changes in HbA1c following transition, five reported no change in HbA1c, one found an improvement in HbA1c for females, and one reported an increase in HbA1c. The remaining two studies, which compared young adults who had transitioned to adult care with those who remained in children's services, found lower HbA1c levels in those who had not transitioned to adult care. In the 10 studies examining transition programmes or interventions, four reported no change in HbA1c between pre- and post-transition and six reported an improvement. Programmes demonstrating improvements incorporated a range of elements, including joint clinics, letters to patients summarizing medical history and support programmes.

Out of 8 observational studies, which assessed the relationship between healthcare transition and clinic attendance, four studies reported less-frequent attendance post-transfer compared with pre-transfer. Change in the frequency of attendance post-transition, where transition programmes or interventions were in place, was assessed in five studies, and all studies found reduced attendance rates post-transition.

In 4 observational studies that examined the relationship between diabetes-related hospitalizations and healthcare transition, no change in hospitalizations were found in 2 studies, one study found an increase in hospitalization frequency, and the remaining study found fewer diabetes-related hospitalisations in those who remained in paediatric care.

VI. RESISTANCE TO NEW PHYSICIANS

Young people with T1DM generally experienced greater difficulties in accessing and keeping their own heath care after transition [19]. They felt that adult clinics were less reachable because of less-frequent appointments and less multi-dimensional support, and had difficulties in building consistent relationship with their new health caretakers. Clinic cancellation rates were higher in adult services [20]. Engagement may have been disturbed by various reasons, for example, the location and opening hours of adult clinics, capacity to travel, other appointments such as work or education, parking difficulties or transport costs and latency time [21]. Patients in adult services were reportedly less likely to be followed up if they did not attend their appointments and were not automatically rebooked [22].

Within adult diabetes clinics, young adults preferred consultations with familiar health care staff [22]. Health care staffs in adult health providers tend to make generalisations and judgements in relation to young adults, which undermined relationship development within the clinic [22]. Fear or perception of being told off or judged by health care professionals for unsatisfactory glycaemic control was a barrier to clinic attendance studies [22]. Young adults felt health care staff did not take the time to understand the struggle they experienced to achieve glycaemic control [23]. Some felt out of place in adult services and that the care was inappropriate for the needs of their age group [21], many were less satisfied with adult care than children's care [22].

VII. Psychological and Socioeconomic Issues

Mental health is a more common concern among young adults with T1DM than those without it, which is usually a miscalculated problem [25]. A recent study found much higher prevalence of psychosocial illness among adolescents with T1DM than those without it [55.95% vs. 20%; P<0.0001] [25]. Major depressive disorder was the most common, while conduct disorder and generalized anxiety disorder were less. Depression and anxiety have been identified as a risk factor for impaired adherence to diabetes care and higher HbA1c [26]. It is important to monitor and refer older adolescents and young adults with T1DM to appropriate mental health resources [12].

Substance use was common among adolescents and young adults with T1DM, as well as those without diabetes. High-risk alcohol use has been reported in 12.9% of 14–19-year old patients with T1DM [27]. The effects of alcohol on glucose and a risk for hypoglycaemia unawareness should be made well known to the young people with T1DM [27]. Among patients with diabetes, tobacco use is an independent modifiable risk factor for development of cardiovascular diseases, diabetic neuropathy, and nephropathy [28]. Illicit drug use has been found to be a risk factor for non-adherence and diabetic ketoacidosis (DKA) [29].

Eating disorders and eating disordered behaviour also are of concern in adolescents with T1DM. Rates of eating disorders among adolescents with T1DM are estimated at 10%; it is twice as high as in girls without diabetes and e the incidence of eating disorders increases into young adulthood [30]. Insulin restriction is the most concerned eating disordered behaviour for adolescents with T1DM, which results in hyperglycaemia creating glucosuria leading to weight loss. This behaviour is very common and could occur in patients who do not meet criteria for an eating disorder [30].

Pregnancy planning is an important topic to be young woman with T1DM. The American Diabetes Association (ADA) recommends preconception planning should be discussed with all women with T1DM starting at puberty [12]

Lack of access to adequate health care coverage remains to be one of the most significant issues to transition for the emerging adult population with T1DM. They may have increased pressure to secure full-time employment to maintain health insurance for on-going medical care for their chronic illness because of changes in legal status and living situation that occur in early adulthood. Due to high co-pays, some may be pushed to switch to less optimal insulin use or become less adhered to diabetes care [31].

VIII. Structured Support Aids for The Adolescent and Young Patients in Morioka Area, Japan

A. *Transition Trial*

1) *The aims of this study*

The first aim of the present study is to provide a seamless transitional process to the young patients with T1DM, considering changes of patients' status, progression of complications, and physical and personality maturity along with age, at a joint clinic operated by both paediatricians and physicians. We also endeavour to ascertain the changes in psychological burden on the patients, quality of life due to transition, self-care behaviours, and blood glucose control during the transition.

B. *Method/ Design*

1) *Design*

A prospective, uncontrolled trial is conducted in Morioka Children's Hospital [paediatric and joint clinics], and Iwate Medical University (an adult clinic). The structured transition program is a multidisciplinary intervention designed to provide additional support in the transition period, which include age-specific education and counselling on basic knowledge on diabetes, sick days, family planning, pregnancy and childbirth, drinking and smoking, and carbohydrate counting, by paediatricians, physicians, nurses, registered dieticians, and clinical psychologists. A web site provides information on the joint clinic by a newsletter, and downloadable forms of useful knowledge for diabetes self-management. Central to the program is use of the flash glucose monitoring system FreeStyle Libre (Abbott Diabetes Care Ltd., Witney, UK) to acquire flexible insulin therapy corresponding to dietary modification, including the carbohydrate counting. Subjects are seen in the paediatric care setting, the joint clinic at the Children's Hospital, for 12 months and then transferred to the adult care setting where they are seen for one year. There will then be a one-year follow-up period for outcome assessment. This study is going to be approved by the Institutional Review Board of Iwate Prefectural University, Iwate Medical University, and Morioka Children's Hospital, respectively. All participants must complete an informed consent at study enrolment. The competitive research funding by Iwate Prefectural University is financially supporting this project.

2) *Participants*

a) *Inclusion criteria*

- Established T1DM diagnosis for a minimum of one year
- Older than 15 years of age
- At least 1 visit during the previous year with the paediatric endocrinologist at Morioka Children's Hospital

- Ability to participate in all aspects of this clinical trial
- Written informed consent/assent must be obtained and documented
- Willingness to participate in the joint clinic and to be transferred to the adult clinic
- Resident of Iwate Prefecture

b) Exclusion criteria

- Pregnant or lactating females or intent to become pregnant during the next 3 years
- Condition(s) which in the opinion of the investigator may interfere with the subject's ability to participate in the study
- Prior enrolment in the current study
- Prior enrolment of a sibling in the current study
- Current participation in another clinical trial or participation in another clinical trial in the 6 months prior to enrolment

3) Study procedures

a) Recruitment

Eligible patients are identified in Morioka Children's Hospital. The investigators introduce the study and provide a document on information to all eligible patients. If the prospective participant agrees to be approached, the investigators make contact during the clinic to answer any questions or concerns regarding the study. If the subject agrees to participate, informed consent is obtained at the time or at the next routine paediatric clinic visit. The researchers make contact during the clinic to answer any questions or concerns regarding the study.

b) Baseline assessment

Once consent has been obtained, the baseline assessment is completed as part of the initial visit. Baseline characteristics collected are: age, gender, level of education, family structure, distance from the treatment centre, smoking and alcohol use, comorbid conditions, concomitant medications, and family history of diabetes. In addition, baseline assessment includes detailed initial medical history, measurement of weight and height, blood pressure, centralized venous A1C, insulin use, clinic attendance, and completion of baseline patient satisfaction questionnaires.

c) Hisotorical measures

- Sociodemographic: age, sex, level of education, persons living with participants/ family structure, and distance from the treating centre
- Medical history: detailed initial medical history; family history of diabetes-related complications, social habits (smoking, alcohol, illicit drug use), follow-up interim history with focus on hospital visits for hypoglycaemia and diabetic ketoacidosis
- Insulin dosage and method of delivery
- Frequency of medical care (retinal, monofilament, lipid profile testing and microalbumin to creatinine ratio)
- Concomitant medications: all longstanding therapies, with the emphasis placed on insulin therapy
- Questionnaires (Diabetes therapy Related QOL; Problem Areas in Diabetes; Summary of Diabetes Self-Care Activities)

d) Physical examination measures

- Anthropometric measurements: height, weight, and BMI
- Blood Pressure
- Systems physical examination: general survey, skin, head, neck, chest, heart, abdomen, musculoskeletal/ extremities, and neurologic (including lower extremity monofilament testing)
- Laboratory measures A1C, lipid profile, creatinine

e) Transition interventions

The structured transition program is a multidisciplinary intervention aiming to provide additional support during the transition period, which include age-specific education and counselling on basic knowledge on diabetes, the differences in the structure of adult and paediatric diabetes care, sick days, family planning, pregnancy and childbirth, drinking and smoking, and carbohydrate counting by paediatricians, physicians, nurses, registered dieticians, and clinical psychologists. Written information is provided. The intervention lasts 24 months, 12 months at a joint clinic in paediatric care and 12 months in adult care. Recommended revisits interval ranges 4-8 weeks. At every visit, laboratory tests are performed. The participants complete the questionnaires at the beginning and the end of the joint clinic, and the end of the adult clinic.

f) Primary outcomes

The primary objectives of the study are:

- To compare the mean A1C levels before and after the transition clinic, or before and after the adult clinic
- To compare the rates of diabetes related emergency room visits and hospitalizations for DKA and hypoglycaemia before and after the transition clinic, or before and after the adult clinic
- To compare the patient satisfaction and perception of the care before and after the transition clinic, or before and after the adult clinic, using self-administered questionnaires

g) Adverse events and safety

Due to the nature of the intervention, it is not expected that serious adverse events related to the intervention will occur. However, adverse events will be collected from the time of signing the Informed Consent. The following adverse events will be recorded in the subject's medical records and on the case report form:

- Any medical occurrences requiring medical intervention
- Any action or outcome (e.g., hospitalization, discontinuation of therapy, etc.) will also be recorded for each adverse event

IX. Discussion

Mental health is a more common concern among young our knowledge, this study is the first project to perform a prospective evaluation of a structured transition program in Japan. This study incorporates the recent recommendation of the American Diabetes Association (ADA) [32], for transition from paediatric to adult care, emphasizing that "Both paediatric and adult diabetes care providers should provide support and links to resources for transitioning young adults."

To date, there are no studies that have directly compared various transition interventions, and there is a limited number of clinical trials on transitional medicine in the T1DM field. In our present project, we follow the transition procedure of randomised controlled trial by Sequeira PA, et al [33], the Let's Empower and Prepare (LEAP) study, which included 51 young patients with T1DM, and successfully facilitated their transition to adult care without a decrease in clinical follow-up. The LEAP program provided had four major components that we inherit in our study, diabetes education tailored to patients' developmental stage, a joint clinic operated by both a paediatrician and a physician, carbohydrate counting education, and a private social networking website to facilitate social support. Indeed, the LEAP study first proved the efficacy of tailored diabetes education and joint clinic in the transition [33]. Carbohydrate counting is a popular method of calculating grams of carbohydrate consumed at meals and snacks, allowing patients with T1DM to adjust the mealtime insulin dose considering the amount of carbohydrates to eat. There is significant evidence that this method improves blood glucose controls and quality of life of the patients [34]. We introduce a modification to this meal planning method with novel a novel flash glucose monitoring system for interstitial glucose measures, which significantly reduced time in hypoglycaemia without deterioration of HbA1c, and improved treatment satisfaction [35]. Internet-based self-management among young patients has found some success [36]. Recent trends indicate a major shift to incorporate mobile telecommunication technologies into health behaviour interventions. These technologies offer several advantages for health behavioural interventions, including information and messages tailored to the participant, quick access, increased cultural sensitivity, and anonymity, which may be attractive regarding sensitive health issues [37].

X. Conclusion

Psychological and socioeconomic issues characteristic of young adults with T1DM similar to those reported in the Caucasian population were found in our study, as well. Continuous and developmentally tailored diabetes medical care by health care providers trying to integrate the patient's life circumstances is central to resolution of these issues. The findings of the current project are expected to support the routine implementation of standardized intervention during the transition period not only in T1DM, but also all other areas of care for emerging adults with chronic medical conditions.

References

[1] M. McPherson, P. Arango, H. Fox, C. Lauver, M. McManus, P. W. Newacheck, J. M. Perrin, J. P. Shonkoff, and B. Strickland, "A new definition of children with special health care needs," Pediatrics, vol.102, pp. 137-40, Jul. 1998.

[2] L. F. Zhang, J. S. Ho, and S. E. Kennedy, "A systematic review of the psychometric properties of transition readiness assessment tools in adolescents with chronic disease," BMC Pediatr, vol. 14, P. 4, Jan. 2014, doi: 10.1186/1471-2431-14-4.

[3] J. Silverstein, G. Klingensmith, K. Copeland , L. Plotnick, F. Kaufman, L. Laffel, L. Deeb, M. Grey, B. Anderson, L. A. Holzmeister, and N. Clark, "Care of children and adolescents with type 1 diabetes: a statement of the American Diabetes Association", Diabetes Care, Vol. 28, pp 186-212, Jan. 2005.

[4] International Diabetes Federation, IDF Diabetes Atlas, 7th ed., Brussels, Belgium: International Diabetes Federation, 2015.

[5] S. S. Jaser, "Psychological problems in adolescents with diabetes," Adolesc Med State Art Rev, vol. 21, pp. 138-51, Apr. 2010.

[6] R. Whittemore, S. Jaser, A. Chao, M. Jang, and M. Grey, "Psychological experience of parents of children with type 1 diabetes: a systematic mixed-studies review," Diabetes Educ,vol. 38, pp. 562-79, Jul.-Aug. 2012.

[7] M. Hirose, E. A. Beverly, and K. Weinger, "Quality of life and technology: impact on children and families with diabetes," Curr Diab Rep, vol. 12, pp. 711-20, Dec. 2012.

[8] K. Hassan, R. Loar, B. J. Anderson, and R. A. Heptulla, "The role of socioeconomic status, depression, quality of life, and glycemic control in type 1 diabetes mellitus," J Pediatr, vol. 149, pp. 526-31, Oct. 2006.

[9] J. Arnett, "Emerging adulthood: a theory of development from late teens through the twenties," Am Psychol, vol. 55, pp. 469–480, May. 2000.

[10] J. Arnett, "Conceptions of the transition to adulthood: perspectives from adolescence through midlife," J Adult Dev, vol. 8, pp. 133–143, Apr. 2001.

[11] L. M. Laffel, L. Vangsness, A. Connell, A. Goebel-Fabbri, D. Butler, and B. J. Anderson, "Impact of ambulatory, family-focused teamwork intervention on glycemic control in youth with type 1 diabetes," J Pediatr, vol. 142, pp. 409–416, Apr. 2003.

[12] A. Peters, L. Laffel, and American Diabetes Association Transitions Working Group, "Diabetes care for emerging adults: recommendations for transition from pediatric to adult diabetes care systems," Diabetes Care, vol. 34, pp. 2477-85, Nov. 2011.

[13] J. Weissberg-Benchell, H. Wolpert, and B. J. Anderson, "Transitioning from pediatric to adult care: a new approach to the post-adolescent young person with type 1 diabetes," Diabetes Care, vol. 30, pp. 2441–2446, Oct. 2007.

[14] M. Nakhla, L. E. Bell, S.Wafa, and K. Dasgupta, "Improving the transition from pediatric to adult diabetes care: the pediatric care provider's perspective in Quebec, Canada," BMJ Open Diabetes Res Care, vol. 5, e0003902017, Jun. 2017, doi: 10.1136/ bmjdrc-2017-000390.

[15] L. Scott, T. M Vallis., M. Charette, A. Murray, and R. Latta, "Transition of care: researching the needs of young adults with type 1 diabetes," Can J Diabetes, vol. 29, pp. 203–210, Jan. 2005.

[16] C. S. Price, S. Corbett, N. Lewis-Barned, J. Morgan, L. E. Oliver, G. Dovey-Pearce, "Implementing a transition pathway in diabetes: a qualitative study of the experiences and suggestions of young people with diabetes," Child Care Health Dev, vol. 37, pp. 852–860, Nov. 2011.

[17] L. Perry, J. M.Lowe, K. S. Steinbeck, and J. S. Dunbabin, "Services doing the best they can: service experiences of young adults with type 1 diabetes mellitus in rural Australia," J Clin Nurs, vol. 21, pp. 1955–1963, Jul. 2012.

[18] A. M. Sheehan, A. E. While, and I. Coyne, "The experiences and impact of transition from child to adult healthcare services for young people with Type 1 diabetes: a systematic review," Diabet Med, vol. 32, pp. 440-58, Apr. 2015.

[19] N. Van Walleghem, C. A. Macdonald, and H. J. Dean, "Evaluation of a systems navigator model for transition from pediatric to adult care for young adults with type 1 diabetes," Diabetes Care, vol. 31, pp. 1529–1530, Aug. 2008.

[20] N. Kime, "Young people with type 1 diabetes and their transition to adult services," Br J Community Nurs, vol. 18, pp. pp. 6–8, Jan. 2013.

[21] P. Rapley P., L. Hart, G. Babel, and J. Kaye, "Young adults transferring to a young adult diabetes clinic: Hba1c and missed appointments," Australian Diabetes Educator, vol. 10, pp. 10–12, 2007.

[22] K. Visentin, T. Koch, and D. Kralik, "Adolescents with Type 1 diabetes: transition between diabetes services," J Clin Nurs, vol. 15,pp. 761–769, Jun. 2006.

[23] S. Brierley, C. Eiser, B. Johnson, V. Young, and S. Heller, "Working with young adults with type 1 diabetes: views of a multidisciplinary care team and implications for service delivery," Diabet Med, vol. 29, pp. 677-681, May. 2012.

[24] R. Snow, and N. Fulop, "Understanding issues associated with attending a young adult diabetes clinic: a case study," Diabet Med,: vol. 29. pp. 257-259, Feb. 2012.

[25] S. Khandelwal, G. S. Sengar, M. Sharma, S. Choudhary, and N. Nagaraj, "Psychosocial Illness in Children with Type 1 Diabetes Mellitus: Prevalence, Pattern and Risk Factors," J Clin Diagn Res, vol. 10, pp. SC05-SC08. Sep. 2016.

[26] M. Herzer, and K. K. Hood, "Anxiety symptoms in adolescents with type 1 diabetes: association with blood glucose monitoring and glycemic control," J Pediatr Psychol, vol. 35, pp. 415-25, May. 2010.

[27] M. A. Knychala, M. L. Jorge, C. K. Muniz, P. N. Faria, and P. T. Jorge, "High-risk alcohol use and anxiety and depression symptoms in adolescents and adults with type 1 diabetes mellitus: a cross-sectional study," Diabetol Metab Syndr, vol. 7, p. 24. Mar. 2015.

[28] D. Haire-Joshu, R. E. Glasgow, and T. L. Tibbs, "Smoking and diabetes," Diabetes Care, vol. 22, pp. 1887-98, Jan. 1999.

[29] E. A. Warner, G. S. Greene, M. S. Buchsbaum, D. S. Cooper, B. E. and Robinson, Diabetic ketoacidosis associated with cocaine use. Arch Intern Med 1998, vol. 158, pp. 1799-802.

[30] R. C. Peveler, K. S. Bryden, H. A. W. Neil, C. G. Fairburn, R. A. Mayou, D. B. Dunger, and H. M. Turner, "The relationship of disordered eating habits and attitudes to clinical outcomes in young adult females with type 1 diabetes," Diabetes Care, vol. 28, pp. 84–88, Jan. 2005.

[31] K. A. McBrien, C.Naugler, N. Ivers, R. G. Weaver, D. Campbell, L. Desveaux, B. R. Hemmelgarn, A. L. Edwards, N. Saad, D. Nicholas, and B.J. Manns, "Barriers to care in patients with diabetes and poor glycemic control—A crosssectional survey," PLoS One, vol. 12, e0176135, May. 2017, doi: 10.1371/ journal.pone. 0176135.

[32] American Diabetes Association, "Children and Adolescents: Standards of Medical Care in Diabetes-2018," Diabetes Care, vol. 41(Suppl 1), pp. S126-S136, doi: 10.2337/dc18-S012.

[33] P. A. Sequeira, E. A. Pyatak, M. J. Weigensberg, C. P. Vigen , J. R. Wood, V . Ruelas, L . Montoya, M . Cohen, H . Speer, S. Clark, and A. L. Peters, "Let's Empower and Prepare (LEAP): Evaluation of a Structured Transition Program for Young Adults With Type 1 Diabetes," Diabetes Care, vol. 38, pp. 1412-9, Aug. 2015, doi: 10.2337/dc14-2577.

[34] DAFNE Study Group, "Training in flexible, intensive insulin management to enable dietary freedom in people with type 1 diabetes: dose adjustment for normal eating (DAFNE) randomised controlled trial," BMJ, vol. 325, pp. 746-7151, Oct. 2002.

[35] P. Oskarsson, R. Antuna, P. Geelhoed-Duijvestijn, J. Kröger, R. Weitgasser, and J. Bolinder, "Impact of flash glucose monitoring on hypoglycaemia in adults with type 1 diabetes managed with multiple daily injection therapy: a pre-specified subgroup analysis of the IMPACT randomised controlled trial," Diabetologia, doi: 10.1007/s00125-017-4527-5, Dec. 2017.

[36] V. van der Meer , M. J. Bakker, W. B. van den Hout , K. F. Rabe, P. J. Sterk, J. Kievit J, W. J. Assendelft, J. K. Sont and SMASHING (Self-Management in Asthma Supported by Hospitals, ICT, Nurses and General Practitioners) Study Group, "Internet-based self-management plus education compared with usual care in asthma: a randomized trial," Ann Intern Med, vo. 151, pp. 110-20, Jul. 2009.

[37] J. S. Huang, L. Terrones, T. Tompane, L. Dillon, M. Pian, M. Gottschalk, G. J. Norman, and L. K. Bartholomew, "Preparing adolescents with chronic disease for transition to adult care: a technology program," Pediatrics, vol. 133, pp. :e1639-46,Jun. 2014, doi:10.1542/peds.2013-2830.

A Study on Facial Expressions when Nursing Students Care for Patients

Natsuko Miura, Kazuma Takahashi
Faculty of Nursing
Iwate Prefectural University
Takizawa, Japan
e-mail: {natsuko, t_kazuma }@iwate-pu.ac.jp

Yoshitosi Murata
Faculty of Software and Information Science
Iwate Prefectural University
Takizawa, Japan
e-mail: y-murata@iwate-pu.ac.jp

***Abstract*— The aim of this study was to clarify what kind of their facial expression nursing students selected; be it dependent on their patients' situations and the nursing care. Furthermore, educational needs concerning facial expressions are examined. In accordance with P. Ekman's facial expression analyses, photographs of 10 types of facial expressions were taken and the emotional components perceived therein analyzed. Using the 10 expressions, several aspects were studied. Twenty-seven students responded to questions. Results revealed students may misunderstand the expression of fear as one of disgust or sadness. Furthermore, they tend to think that the expressions of surprise, fear, disgust, anger, and sadness are inappropriate facial expressions when talking to a patient or conducting a physical assessment. Twenty students responded to educational needs regarding facial expressions. Their remarks included "lessons on how to make facial expressions corresponding to situations in simulations and actual practices are desirable" and "effectiveness of facial expressions." In addition, having the students' own facial expressions mechanically analyzed or evaluated by others may offer them an opportunity to learn how to express emotions appropriately. As these aspects are in accordance with the students' educational needs, an early introduction to the education on facial expressions is desirable.**

Keywords-facial expression; nursing student; physical assessment; communication.

I. INTRODUCTION

Because it is a requirement for nurses to understand their patients' needs and respond accurately, proficient communication skills are indispensable for their profession [1]. However, in the actual education practices for nurses in Japan, a deterioration in the nursing students' ability to sympathize as well as other interpersonal abilities has been evident; thus, it is difficult for students to build a mutual understanding with their patients through their nursing practices [2]. In addition, a survey on university students' communication skills from the perspective of social skills revealed that the emotional sensibility of nursing students, which is required to understand nonverbal information, is poorer than that of students who have specialized in care for people who have disabilities. Furthermore, their ability to control their emotions, which is required to keep their emotional sensibility in check, is higher than that of pedagogic students [3]. Although the impact of the facial expressions of nurses [4], physical therapists [5], and pharmacists [6] on their patients has been reported, no study has revealed how nursing students choose their facial expressions; be it based on their patients' situations or events. There has also been no clear indication on the need for education with reference to facial expressions.

In view of these circumstances, the aim of the present study was to clarify what kind of their facial expression nursing students selected; be it dependent on their patients' situations and/or nursing events. Furthermore, educational needs with reference to facial expressions are examined.

In Section 2, the method employed in the study is explained. In Section 3, the results of the feelings analysis of 10 kinds of expressions, which were developed for this investigation, are outlined. In addition, facial expressions when a nursing student provides care for a patient are described. A brief discussion is offered in Section 4; conclusions and recommendations for future research are presented in Section 5.

II. METHOD

In accordance with P. Ekman's facial expression analyses [7], photographs of 10 types of facial expressions were taken; these are depicted in in Figure 1 (A: blank expression; B-E: four levels of happiness; F: surprise; G: fear; H: disgust; I: anger; J: sadness). The emotional components perceived therein were analyzed. The Emotion API of Microsoft Cognitive Services [8] [9] was employed to analyze facial expressions. The reason for a number of facial expressions showing happiness is nursing students use different degrees of smiling when caring for patients. The degrees of expressing happiness were derived from the researchers' experience.

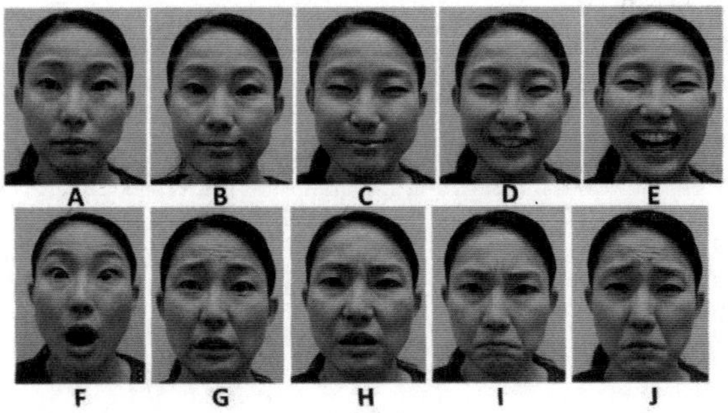

Figure 1. The 10 types of facial expressions: A: blank expression; B-E: four levels of happiness; F: surprise; G: fear; H: disgust; I: anger; J: sadness

Using the 10 expressions, the items thus listed were studied. To assess their adequacy, the facial expressions were rated by means of seven levels; Level 1 denoted "not adequate at all," while Level 7 signified it was "very adequate." The items are as follows:

- How nursing students recognize emotions behind the 10 facial expressions;
- Adequacy of facial expressions when talking with a patient by showing a smile, blank expression, worry, agony or anger;
- Adequacy of facial expressions when conducting a physical assessment of a patient by showing a smile, blank expression, worry, agony or anger;
- Adequacy of facial expressions when communicating a problematic or non-problematic test result.
- Past experience in facial expression training and the emotion felt on that occasion; and
- Educational needs regarding facial expressions.

III. RESULT

A. The analysis of the 10 facial expressions

The analysis of the 10 facial expressions prepared for this study mainly detected components associated with a lack of emotion in Expressions A and B, happiness in C, D and E, surprise in F, sadness in G, I and J, and a lack of emotion in H (Table I). The expressions of fear, disgust, and anger included a great deal sorrow and expressionless components.

TABLE I. THE ANALYSIS OF THE 10 FACIAL EXPRESSIONS EMPLOYING THE EMOTION API OF MICROSOFT COGNITIVE SERVICES

Emotions expressed [a]		Analysis items of emotions [b]							
		Anger	Contempt	Disgust	Fear	Happiness	Sadness	Surprise	Neutral
A	Neutral	0.00	0.00	0.00	0.00	0.00	0.18	0.00	0.82
B	Happiness	0.00	0.01	0.00	0.00	0.07	0.00	0.00	0.92
C	Happiness	0.00	0.00	0.00	0.00	0.84	0.00	0.00	0.16
D	Happiness	0.00	0.00	0.00	0.00	1.00	0.00	0.00	0.00
E	Happiness	0.00	0.00	0.00	0.00	1.00	0.00	0.00	0.00
F	Surprise	0.00	0.00	0.00	0.00	0.00	0.00	1.00	0.00
G	Fear	0.00	0.00	0.02	0.05	0.00	0.77	0.02	0.14
H	disgust	0.02	0.01	0.04	0.01	0.00	0.40	0.01	0.51
I	anger	0.01	0.00	0.00	0.00	0.00	0.93	0.00	0.06
J	sadness	0.00	0.00	0.00	0.00	0.00	0.99	0.00	0.01

a. Emotions expressed by facial expression.
b. Total of 8 items sums to 1. Because it rounds off, the total may not be 1 in some cases.

B. Emotions that nursing students feel from 10 facial expressions

Six 1st-year students, twelve 2nd-year students, and nine 4th-year students responded to the questions. With reference to Expression A, the majority, namely, 81.5% identified it as "others." Of these, 15 students chose "indifference and lack of emotion." Regarding Expression B, 51.9% identified it as "others" while 40.7% associated it with "happiness." Other responses included "smile" and "tense smile." Expression C was linked to "happiness" by 85.2% of the respondents. Expressions D and E were perceived by 92.6% of the students as "happiness." Other responses included expressions that indicated fun and a lack of animosity against someone else. Expression F was associated by 100% of the respondents with "surprise." Expression G was identified with disgust, fear, and sadness by 44.4%, 33.3% and 14.8%, respectively. Expression H was associated with disgust by 88.9% of the respondents. In Expression I, 92.6% perceived anger while 92.6% perceived sadness in Expression J.

C. Facial expressions that nursing students select for communication and care

When caring for patients who displayed various facial expressions, the expression selected by the nursing student is presented in Tables II and III. To assess their adequacy, the expressions were rated by means of seven levels; Level 1 signified "not adequate at all," while Level 7 meant "very adequate." The extent to which nursing students thought their expressions were appropriate is indicated in the table by the average value. Nursing students should consider the following when speaking with patients who have various expressions.

TABLE II. APPROPRIATENESS OF FACIAL EXPRESSION WHEN TALKING

Facial expressions of nurse		Facial expressions of patients				
		Smiling	*Expressionless*	*Worried face*	*Suffering*	*Angry*
A	Neutral	2.3	3.8	3.3	3.4	4.6
B	Happiness	4.3	5.3	4.7	3.6	3.6
C	Happiness	6.2	5.9	4.6	2.9	3.0
D	Happiness	6.6	5.2	3.9	2.6	2.0
E	Happiness	5.9	3.7	2.4	1.9	1.7
F	Surprise	2.6	2.4	2.1	2.0	2.3
G	Fear	1.6	1.9	2.1	2.9	2.7
H	disgust	1.3	1.6	1.5	1.7	1.8
I	anger	1.4	1.5	1.5	1.9	1.9
J	sadness	1.5	1.8	2.9	3.7	3.1

Nursing students chose a facial expression that they consider appropriate for patients expressing a certain emotion. The average value of the 7-level Likert scale is shown.

The students' tendency when talking to a patient with a beaming smile was to "beam in the same way" or "not to overly smile." When a patient had a blank expression, the students tried to "smile," "make a kind smile because the patient might be worried or nervous," or "change facial expressions along with the conversation." When a patient had a worried expression, the students focused on "keeping a worried face in the same way" or "smiling." When the patient displayed a painful expression, the students tried to "keeping a worried face" or "a smile that keeps the patient from feeling pain." When the patient displayed an angry expression, the students tried to have "a serious look" and "not an angry face," but "smile."

TABLE III. APPROPRIATENESS OF FACIAL EXPRESSION WHEN UNDERGOING A PHYSICAL ASSESSMENT

Facial expressions of nurse		Facial expressions of patients				
		Smiling	*Expressionless*	*Worried face*	*Suffering*	*Angry*
A	Neutral	3.4	3.9	3.3	3.4	3.9
B	Happiness	4.7	5.5	5.0	3.9	3.6
C	Happiness	6.1	6.2	5.3	3.6	3.4
D	Happiness	5.9	5.4	4.3	2.9	2.9
E	Happiness	5.0	3.6	2.8	1.9	1.9
F	Surprise	2.0	1.9	1.8	1.8	1.8
G	Fear	1.5	1.7	2.0	2.8	2.4
H	disgust	1.3	1.4	1.3	1.6	1.7
I	anger	1.3	1.4	1.3	1.7	1.9
J	sadness	1.5	1.6	2.0	2.9	2.6

Nursing students chose a facial expression that they consider appropriate for patients expressing a certain emotion. The average value of the 7-level Likert scale is shown

The facial expressions that nursing students chose when conducting a physical assessment of patients are as follows. The students' tendency when talking to a patient with a beaming smile was to "beam in the same way" or "seriously." When a patient had a blank expression, the students tried to "smile," and display "a facial expression that relieves the patient," or a "gentle expression." The students focused on "keeping a worried face in the same way" or "a facial expression that relieves the patient when the patient had a worried expression. When the patient's facial expression was one of pain, the students tried to "keep a worried face" or "a facial expression that relieves the patient," or "expressions in which seriousness is conveyed to patients." When a patient had an angry expression, the students tried to have "a serious look," "not an angry face," and "a facial expression that conveyed worry."

D. *Facial expression when conveying test results*

The facial expressions when conveying test results are presented in Table IV. When the students informed a patient that there was no problem, they chose to wear a "smile," "bright expression" or "assuring expression." When they informed a patient that there was a problem, they opted to have a "serious expression," "smile," or "nervous expression."

E. *The degree of self-confidence of students who make appropriate facial expressions, and learning needs*

The level of confidence the nursing students had in making an adequate expression was 4.5. The averages per year were 4.6, 3.9, and 5.3 among the 1st-year, 2nd year, and 3rd year students, respectively. There were seven students in the school who received training on facial expressions. The level of satisfaction they experienced from the training was 6.4, their level of delight was 4.5, and level of sadness was 1.9.

TABLE IV. THE FACIAL EXPRESSIONS WHEN CONVEYING TEST RESULTS

Facial expressions of nurse		*No problem*	*Have a problem*
A	Neutral	3.1	4.7
B	Happiness	5.1	4.2
C	Happiness	6.4	2.8
D	Happiness	5.9	2.1
E	Happiness	4.0	1.4
F	Surprise	1.6	1.6
G	Fear	1.4	2.2
H	disgust	1.3	1.7
I	anger	1.3	1.7
J	sadness	1.3	2.6

Nursing students chose a facial expression that they consider appropriate for patients expressing a certain emotion. The average value of the 7-level Likert scale is shown.

F. *The degree of self-confidence of students who make appropriate facial expressions, and learning needs*

The level of confidence the nursing students had in making an adequate expression was 4.5. The averages per year were 4.6, 3.9, and 5.3 among the 1st-year, 2nd year, and 3rd year students, respectively. There were seven students in the school who received training on facial expressions. The level of satisfaction they experienced from the training was 6.4, their level of delight was 4.5, and level of sadness was 1.9.

Three students received instructions on facial expressions during their clinical practice from someone other than a teacher. The level of satisfaction they experienced from the instruction was 7.0, their level of delight was 5.0, and sadness, 3.7.

Twenty students responded on educational needs regarding facial expressions. Their remarks included: "lessons on how to make facial expressions corresponding to situations in simulations and actual practices is desirable," "effectiveness of facial expressions," "mutual peer advice because you cannot exactly know what expression you wear unless the other person tells you," "for the purpose of staying closer to patients' feelings, it would be more useful if the education focuses on how to express or perceive emotions instead of making facial expressions," and "it would be great to have an opportunity to teach junior nurses and students the effective facial expressions when dealing with angry patients, emotionless patients, or patients in agony."

IV. DISCUSSION

Most students perceived happiness in Expressions C and D. While their impressions from Expressions H (disgust) and I (anger) were different from the mechanical analysis results, the students' evaluation seemed consistent. On the other hand, opinions were divided for Expression G (fear) both in the analyses and among students; this suggested the possibility of mistakenly being perceived as an expression of disgust or sadness. In a survey targeting nurses, the highest mean accuracy rate of recognizing facial emotion was happiness (99.14%) while the lowest accurately recognized facial expression was fear (47.71%) [10]. Students, like nurses, are expected to be able to assess the feelings of happiness accurately. The expression of happiness is considered to be a facial expression that students can understand. The reason for this is that the facial expression of happiness is different from other facial expressions; the footprint of the crow is formed in the corner of the eye and the corners of the mouth are raised.

Nursing students are confused by the expressions of fear and disgust. The reason is that the two facial expressions are similar with regard to wrinkles between the nose and eyebrows. Psychiatrists who treat adults were significantly better at recognizing fearful and disgusted facial emotions than child-adolescent psychiatrists while the latter were better at recognizing angry facial emotions [11]. Even qualified medical personnel suggest that there is a difference in the ability to recognize negative facial expressions. Training to recognize facial expressions is necessary for students. This has already been developed as the Micro Expression Training Tool. However, it is reported that the training effect is low when communication ability is poor [12]. While facial expressions are often judged in combination with other circumstances, this study may provide students with an opportunity to realize that patients' feelings can be misunderstood. Furthermore, having students' own facial expressions mechanically analyzed or evaluated by others may offer an opportunity for them to learn how to express emotions. As these aspects match the students' educational needs, an early introduction of the education on facial expression is desirable.

In having conversations with or conducting physical assessments on patients who display various expressions, such as smile, worry, agony, and anger, it was revealed that students consider it inadequate to have an expression of surprise, fear, disgust, anger or sadness. Furthermore, they responded that a modest smile could be applied to any situation. However, because the students do not have many opportunities to deal with a patient in agony or anger in a nursing practice, their responses may be a result of their lack of experience. This suggests the necessity to create opportunities for students to experience how to deal with patients in agony or anger through simulation education. In communicating a problematic test result to a patient, the students considered it better to have a blank expression or modest smile. This study did not conduct any survey on what kind of facial expressions patients prefer or what kind of expressions skilled nurses consider adequate; ideally, these themes should be studied in the future.

V. CONCLUSION

This study aims to clarify nursing students choose what kind of their facial expressions depending on their patients' situations and nursing occasions. The results of this study indicate facial expressions that nursing students select when they are with patients. Nursing students thought that a moderate smile could be applied to any situation. However, patients and nurses need to evaluate the actual students' facial expressions.

Currently, there are no systems that students can self-learn about their facial expressions. Therefore, our future work is to develop a system that integrates facial expression analysis system and others evaluation, so that nursing students can learn about communication.

ACKNOWLEDGMENT

The authors would like to thank K. Yoshida for technical assistance with the experiments. This work was supported by Iwate Monozukuri and Software Integration Technology Center (i-MOS) research expenses.

REFERENCES

[1] L. Kourkouta, and I. V. Papathanasiou, "Communication in Nursing Practice," Materia Socio-Medica, vol. 26, no. 1, pp. 65-67, 2014.

[2] F. Ishimitsu, T. Fruya, and M. Hayashi, "Changes in social skills before and after clinical practice over a half year of nursing college students," (in Japanese), Mejiro journal of health care sciences, vol. 5, pp. 61-66, 2012.

[3] Y. Nishizawa, T. Abe, S. Kudo, K. Hanada, and A. Kasai, "A Study of Social Skills in Young Women : Analysis from Social Skills Inventory," (in Japanese), Japanese journal of nursing research, vol. 25, no. 2, pp. 249-259, 2002.

[4] M. Mastumoto, Y. Tawara, K. Hamai, A. Saito, and T. Simatani, "Influences of interventions with different nurses' expressions for relaxation in healthy volunteers: Changes in vital signs and salivary alpha amylase activity levels," (in Japanese), Nursing journal of Hiroshima International University, vol. 10, no. 1, pp. 15-26, 2013.

[5] K. Yamashita, *et al.*, "The effects of the differences between friendly and unfriendly communication on brain activity," (in Japanese), Japanese journal of health promotion and physical therapy, vol. 7, no. 1, pp. 29-34, 2017.

[6] K. Keiko, H. Akiko, Y. Katsunori, and F. Noriko, "Whether Visual Information Loss of Expression by Community Pharmacist Masks to Effect a Change in Feeling of Trust among Help-seeker?," (in Japanese), Yakugaku Zassi, vol. 130, no. 10, pp. 1401-1413, 2016.

[7] P. Ekman, and W. V. Friesen, *Unmasking the face: A Guide to Recognizing Emotions from Facial Clues*. Prentice Hall Direct, 1987.

[8] Y. Murata, K. Yoshida, and N. Miura, "Quantitative Analysis of Communication Ability with Facial Expression Required for Nurses " 32nd SIP SYMPOSIUM, 2017.

[9] *Microsoft Emotion API*. [Online]. Available from: https://azure.microsoft.com/ja-jp/services/cognitive-services/emotion/ 2017.11.08

[10] G. Gultekin, *et al.*, "Facial emotion recognition ability: psychiatry nurses versus nurses from other departments," Clinical and investigative medicine. Medecine clinique et experimentale, vol. 39, no. 6, pp. 61-65, 2016.

[11] M. Dalkıran, *et al.*, "Facial emotion recognition in psychiatrists and influences of their therapeutic identification on that ability," Comprehensive Psychiatry, vol. 69, pp. 30-35, 2016.

[12] J. Endres and A. Laidlaw, "Micro-expression recognition training in medical students: a pilot study," BMC medical education, vol. 9, no. 47, Jul 20 2009. doi:10.1186/1472-6920-9-47

Identification of User Requirements for Assistive Technology for Support of Upper Limb Daily Life Interactions from Stroke and Duchenne Muscular Dystrophy Patients

Anne L van Ommeren[1,2], Mario Iodice[3,4], Marion Main[3], Jackie M Pitchforth[3,4], Franscesco M Muntoni[3,4], Gerdienke B Prange-Lasonder[1,2], Johan S Rietman[1,2], Jaap H Buurke[1,2]

[1]Roessingh Research and Development
Enschede, the Netherlands

[2]University of Twente
Enschede, the Netherlands

[3]Great Ormond Street Hospital
London, the United Kingdom

[4]University College London
London, the United Kingdom

Email: a.vanommeren@rrd.nl, mario.iodice@ucl.ac.uk, marion.main@gosh.nhs.uk, j.pitchforth@ucl.ac.uk, f.muntoni@ucl.ac.uk, g.prange@rrd.nl, j.s.rietman@rrd.nl, j.buurke@rrd.nl

Abstract— **The uptake of assistive technology to improve the functionality of the upper limb in people with disorders affecting the neuromuscular system, such as stroke and Duchenne Muscular Dystrophy, is often limited by a gap between the users' needs and the design of the technology. This study aims to identify the technology-specific end-users' requirements for the development of upper limb assistive technology to support daily life activities, and thereby supporting self-management, in stroke and Duchenne Muscular Dystrophy, based on the results of disease-specific focus groups and specialist consultation. The focus group results showed that: unobtrusive support, intuitive use, and adaptiveness to an individual and disease progression are key for both stroke and Duchenne Muscular Dystrophy patients. The technology-specific end-user requirements identified in this study can be used to supplement general user requirements identified in the literature, in order to improve the design of assistive technology for support of upper limb daily activities.**

Keywords- DMD; Stroke; Assistive Technology; User Requirements; Upper Extremity.

I. INTRODUCTION

People with disorders that affect the neuromuscular system, such as Duchenne Muscular Dystrophy (DMD) and stroke, often suffer from difficulties in performing activities of daily living (ADL) due to reduced functionality of the upper limb (UL) [1]-[3]. Decreased functionality of the UL leads to a decrease in independence and impacts quality of life [4][5].

The demand for technological solutions, which can support or compensate for loss of functionality in motor function, increases with reduced level of independence and UL function [2][6][7]. Nowadays, numerous technological solutions are available ranging from simple assistive tools (e.g., adapted cutlery) to robots that entirely substitute human movements in very severe cases [7]-[9]. In theory, personal assistance or (in)formal care can be reduced by 30-42% through use of assistive technology [8][9]. Unfortunately, the preferences and needs of end-users and their environment are often not met in the technical design of the device which results in many users abandoning these devices [6][10]-[14]. In order to bridge the translational gap between the users' needs and the design of the technology, a user-centred design needs to be used in the development of the UL assistive technology [15]. Input from end-users from the beginning of the design, as done in a user-centred design, is regarded as effective to enhance the chance for uptake [16].

In the literature, focus is placed on general end-user perspectives for UL assistive technology. If assistive technologies are to be used to support independence during daily life activities, they need to be simple to apply [13]-[15][19], easy to use [11][12][17][18], safe [12], pleasurable [17][19][20], of reasonable cost [17][19][21][22], motivating and should be able to provide feedback [13]. The time taken to prepare, set up and maintain assistive devices are key issues for all stakeholders [11][12][17][19]. For stroke patients and carers the device needs to be easy to get on and off a weak, contracted hand/arm as well as intuitive in terms of correctly positioning the device [11][17]. Also, concerns about devices which are time consuming to clean and difficult to store are expressed [11][17]. The appearance of the device is not seen as important factor for either stroke patients nor healthcare professionals [12]. Conversely, for DMD patients, having a mobile device which is also attractive in appearance is an important issue because of the age related social needs. Thus, for self-management it is critical to incorporate the above mentioned features in the design of a device.

Nowadays, designers are focusing more and more on innovative and technically complex assistive technology, and a user-centred design is increasingly adopted. The People, Activities, Contexts and Technologies (PACT) framework was invented to cover all aspects of user-centred design including social and technological aspects. Although, eliciting end-user input through analysis of the PACT aspects is considered as a useful starting point for design

[23], the technological aspect is often neglected. In stroke and DMD populations, gathered end-user-input is often still targeted at rather generic information about the envisioned use of the assistive technology, for instance as reflected by the People, Activity and Context domains of the PACT framework [11][12][20][23]. There is little emphasis on end-users' views towards specific technical aspects, such as intention detection, options for support, and feedback.

The eNHANCE project aims to assist people with DMD and stroke in performing UL daily life activities with the environment. The focus of this project is on innovative aspects of the technology such as intention detection, performance assessment, and behavioural modelling. Therefore, the aim of this paper is to identify end-user requirements, specifically addressing the technical features, for the development of UL assistive technology. This project used questionnaires and focus groups in two target groups, namely stroke and DMD to determine these end-user requirements. The results are outlined in Section III and will be discussed in Section IV.

II. Methods

To elicit user input, the PACT framework [23] was used to design questions to be addressed during disease-specific questionnaires and focus groups with patients, carers and clinicians. To determine the starting point for the technology-specific questions to be discussed during the focus groups, a questionnaire addressing the People, Activity and Context domains of the PACT framework was set up.

A. *Questionnaire*

In order to develop the questionnaire, a literature survey was performed in stroke, to determine the existing body of knowledge regarding user input for UL supporting assistive technologies. The design of the questionnaire was based on published literature on questionnaire design [24][25]. Questions relating to the People (patient characteristics, technological affinity and hand function), Activity (Usage of hand, and which preferred activities) and Context (Table I) were addressed in the questionnaire. Although no paper-based questionnaire was performed in DMD, questions with regard to the People, Activity and Context domains were asked during the DMD focus group.

TABLE I. Questionnaire Statements, Answered on a 5-Point Likert Scale (From Strongly Disagree To Strongly Agree)

The device must be wearable
Storage of the device must be easy
The device must be lightweight
I want to don and doff the device myself
I would lack the confidence to use assistive technology at home
Support of my caregiver is important to use assistive technology
I want to wear the device under my clothes

In order to ensure consistency in the analysis of the user requirements, the design was kept as similar as possible for DMD as for stroke. Yet, the questions were modified for the DMD population, tailoring it to their specific pediatric needs. Further information was sourced from experience in previous related projects (e.g., FlexTension) and DMD healthcare professionals.

B. *Focus Groups*

Two focus groups, one with stroke survivors and carers, and the other with boys with DMD, aged between 15 and 16 years old, and their parents were held in July 2015. Stroke participants were recruited from the Roessingh Rehabilitation Centre, the Netherlands, and the focus group for DMD was publicised in the UK by the Action Duchenne DMD Advocacy group called "Taking Charge". During the focus groups more specific information with regard to the technology was represented by the following themes: support options, intention detection, personalization, feedback and motivational aspects (Table II). Several technical and biomedical experts were present to provide feedback about the use of the state of the art technology in the robotic arm.

TABLE II. Main Questions Discussed During Focus Groups

Introduction project, participants and focus group	
Support options	
- How would you like to be supported by the system? 1. System takes over entire movement 2. System compensates for gravity 3. System supports as needed	
Intention detection	
- Can you imagine in which way a system could detect your movement intention? Which of the following options would you find best, acceptable and unacceptable:	
A. Subconscious 1. Eye-tracking 2. Sensors (movement/pressure/force) 3. Muscle activation	**B. Conscious** 1. Voice recognition 2. Joystick 3. Pushing a button
Personalization	
- Would you like to have a system that can adapt itself to your personal preferences? - Can you imagine examples of how such a system could be personalized? - Would you like the system to detect the activity you are performing? - How much time is acceptable for the system to get used to your preferences?	
Feedback and motivation	
- Would you like to receive feedback from the system? If so, about 'how well' and/or 'how much' you performed? - What kind of feedback would you like to receive? Audio, visual, graphs and tables or vibrotactile? - Would you like to be encouraged by the system to use your upper limb?	

The focus group with DMD took place in the United Kingdom (UK), while the focus group with stroke survivors took place in the Netherlands. Prior to the start of the study, written informed consent, and agreement for audio-recording of the focus group was obtained from all participants and if needed in case of DMD, their parents. The DMD focus group was conducted as part of a larger project run by the Action Duchenne DMD Advocacy group called "Taking Charge" which obtained consent for videoing and recording. No ethical approval was required for focus groups in this context. In the Netherlands, ethical clearance was obtained from the medical ethical committee Twente, Enschede, the Netherlands, in May 2015.

1) Stroke Focus Group

All topics were accompanied by examples (visual where possible) so that the participants could envisage the options more easily. Using an interactive presentation, those items were put up for discussion by asking the participants input via a variety of user interactions, such as voting and ranking, combined with plenary discussions between all participants about their thoughts, ideas, opinions, experiences and expectations.

2) DMD Focus Group

The format of the DMD focus group differed slightly from the stroke group, to accommodate expression of individual opinions in this group of younger participants. Following preliminary introductions and completion of a baseline questionnaire, an interactive presentation on the eNHANCE project was given. Once the presentation was completed, the end-users gathered together as a group to discuss the presented questions without the presence of the technical and medical experts. This was done to promote participation from young adults who may not voice opinions in the presence of adult experts. The group answers were then presented to the technical and medical attendees. Group ideas and suggestions were recorded by Dictaphone and video for later analysis.

C. User Requirement Identification

The qualitative data coming from audio-recordings and notes from the focus groups was elaborated. Transcripts were discussed between researchers and direct comments were subsequently grouped together. From this, common topics were identified to describe the user perspectives per predefined theme. Thereafter, user requirements were compiled according to preferences expressed by the majority of the participants in each focus group. Subsequently, user requirements were discussed between clinical experts for DMD and stroke separately, involving rehabilitation physicians, physiotherapists and clinical researchers. All requirements were independently prioritized by at least three clinical experts per target population using the MoSCoW method; Must have (M), Should have (S), Could have (C), Won't have (W) [26]. The MoSCoW method is a technique used for prioritization of requirements with stakeholders to highlight the importance placed on each requirement. The final priority was based on most votes given for the corresponding user requirement.

III. RESULTS

In the result section, the findings from the questionnaire in stroke and the end-user requirements as derived from the focus groups in stroke and DMD are presented.

A. Findings from the Questionnaire in Stroke

Findings from the questionnaire in stroke from the People Activity and Context domains of the PACT framework are presented separately.

1) People Domain Findings from the Questionnaire

In total, seven stroke survivors filled out the questionnaire. Most of the questionnaire respondents were male, above the age of 60 and in the chronic phase after stroke (Table III). Main problems in functional use of the hand and arm were a lack of fine motor skills and control of the hand.

2) Activity Domain Findings from the Questionnaire

The majority of the respondents used their affected hand and arm at least sometimes. Dressing and undressing, biking and using the affected hand as supporting hand were the activities in which the most respondents used their affected hand and arm. With regard to activities in which they would like to use their affected arm; domestic chores, eating, drinking and cooking and dressing/grooming were reported most often. Personal hygiene, outdoor activities, mobility but also fine motor skill activities and hobbies were mentioned.

3) Context Domain Findings from the Questionnaire

Findings answering the questions of the 'Context' domain of the PACT framework are summarized in Figure 1. The majority of the respondents either agreed or strongly agreed that assistive technology should be easy to don and doff, wearable and light weight, and easy to store. All participants would be confident to use assistive technology independent at home. However, opinions about the amount of support needed from the caregiver and whether the device should be worn under the clothes or not were more diversified.

TABLE III. PRIMARY END-USER CHARACTERISTICS

Table Head	**Table Column Head**				
	Target population	***% Male***	***Mean age (range)***	***% Wheelchair user?***	***Stroke: Onset since disease in years (range)***
Questionnaire	Stroke (n=7)	71%	66 (45-78)	0%	3.4 (2.5-4.5)
Focus group	Stroke (n=3)	100%	70 (67-75)	0%	3.3 (2-4.5)
	DMD (n=6)	100%	15.2 (15-16)	100%	Not Applicable

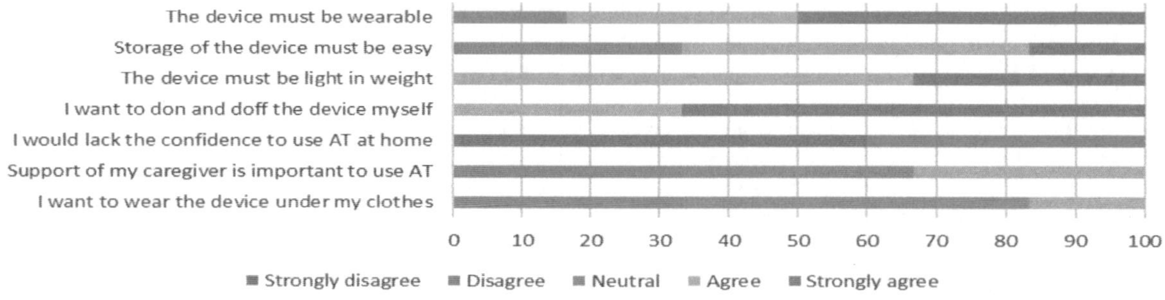

Figure 1. Perceptions of stroke survivors on contextual questions with regard to assistive technology, expressed in percentages

B. *Focus Group Derived End-user Requirements Focused on Technological Aspects*

Nine primary end-users (DMD boys and stroke survivors) and eleven secondary end-users (partners and parents) were included in the focus groups. Patient characteristics can be found in Table III.

TABLE IV. USER REQUIREMENTS WITH CORRESPONDING PROIRITY. M = MUST HAVE, S = SHOULD HAVE, AND '-'= NOT RRIORITIZED OR NOT APPLICABLE

User requirement	Priority	
	DMD	***Stroke***
Support options		
The system must allow the user to move actively without replacing human function	-	M
The amount of the support must consider the existing contractures on the upper limb	M	M
The system must support arm function during reaching and fine motor control of the hand	-	M
Intention detection		
The system must subconsciously detect the intention of the user	M	S
The system must have the possibility to switch to conscious control of the user	S	S
Personalization		
The system must be adaptable to personal limitations and needs	M	M
The system must be able to learn user preferences within one week	-	S
The system must be adaptable to different kind of movements in tasks related to activities of daily life	-	M
The system must consider the eventual users' deteriorating condition overuse, pain and muscle deterioration	M	-
Feedback and motivation		
The system must provide feedback during the daily tasks (quality and quantity of movement)	M	-
The system must not overproduce feedback	S	M
The system must keep track of performance based on personal capacity and should use that to motivate the user	-	M
The system must motivate the user to be active during the movements	S	M

In Table IV, a list of user requirements resulting from the translation of user expressions from end-users gathered during the focus groups can be found. In summary, both groups favoured assistance only as needed. The DMD group favoured subconscious control (e.g., eye-tracking, movement sensors, muscle activation) with an option to switch off the subconscious control. The stroke group was more divided with some favouring conscious (e.g., voice, joystick, button pressing) over subconscious control. Both groups wished for personalization of the system but differed with regard to individual characteristics, which should take into account mainly disease-specific aspects related to nature and severity of motor and/or cognitive limitations, as a basis for personalization.

IV. DISCUSSION

Reduced function of the upper limbs in both people with DMD and stroke impacts their functional independence and quality of life [1]. Although the use of assistive technology is promising, user-centred design methods are needed to include end-users in the design process and to enable development of devices that better suit the needs of the users [15]. Findings from the questionnaire and the user requirements gained from focus groups in the two selected target populations, DMD and stroke, provided useful and specific information on the technological features of assistive technologies for the UL from an end-users' perspective.

Remarkably, the themes and related topics identified by end-users as being essential requirements for the design of an intelligent UL supporting assistive technology were comparable between DMD and stroke participants. Despite these similarities, there were substantial differences between specific requirements between populations for each of those themes (see Table IV for examples) attributable to differences in the two populations (Table V). Unlike the DMD population, both men and women are affected by stroke, yet only men participated in the focus group. However based on findings of former focus groups including both women and men, no different outcomes are expected when women would have been included [15][27].

The functional benefit of a device must be balanced with its burden of use [7]. Minimization of the burden must result from including end-users in the design of the device. In addition to the end-user requirements about technological aspects of assistive technology gained from this study, there are a number of other aspects that are important for the adoption of assistive technology. In line with the findings from the questionnaire performed in this study, assistive technologies need to be: easy to use, portable, safe, and easy to don and doff during daily activities such as eating/drinking, preparing food, personal hygiene, as well as supporting hand in stroke in order to gain more independence and perform desirable activities [11][12][15][17][18][20][22][27].

As stroke survivors prefer to use assistive technology at home, storage of assistive technology must be easy. In DMD, assistive technology must be able to be attached to the wheelchair and to be used during the entire day [19], whereas in stroke an assistive technology can preferably be used several times a day depending on the easiness with which the system can be put on by the patient.

In stroke, accessibility or knowledge about the device is also identified as important factor for the uptake of UL assistive technology [28]. In general, the results of the questionnaire of this study and previous research incorporating input from both primary and secondary end-users are consistent with the 17 design and engineering criteria as set up by Batavia et al. [29]. Those criteria are applicable in both stroke and DMD.

TABLE V. PRIMARY DIFFERENCES BETWEEN TARGET POPULATIONS

Stroke	DMD
Static	Progressive
Male and female	Male
Unilateral involvement	Bilateral involvement
Predominantly older adults	Difficulties begin in early teens
Increased muscle tone (spasticity)	Spasticity not involved
Pain	Pain rarely present
Generally ambulant	Non-ambulant

Although previous research has identified a need for feedback, (mechanical) adjustment to patients and the ease of use, there has not been a specific focus on the technological aspects of assistive technology [11][12][22]. During the focus groups, almost every participant, both stroke and DMD, pointed out that it is important that their own existing power, movement and function must be enhanced, rather than replaced, by the system in order to be as independent as possible. Within both populations, highly individual aspects such as variety in disease severity, in addition to personal preferences and interests, needs to be taken into account during the personalization of interfaces. Furthermore, the differences between target populations as highlighted in Table V, resulted in different requirements. Whereas stroke survivors stated they always want to regain more functionality, regardless of the severity, in DMD, the primary concern is to fight deterioration at every stage of the disease. Therefore, it is of great importance that a system can be personalized not only to the personal preferences and interests of the user, but also to disease-specific needs in the motor and cognitive domains [28].

With regard to detection of the movement intention of the user, our participants predominantly preferred sub-conscious control, with the possibility to switch to or combine it with conscious control. In DMD, eye-tracking was favoured by all the boys and their parents. Although subconscious was most preferred in stroke, there was difference between participants in the preferred option (e.g., movement sensors, eye-tracking, muscle activation).

In order to reduce or reverse functional decline in motor function, active engagement during movement and intensive use of the arm of hand are crucial [30]. Although stroke survivors do not feel the urge to receive direct feedback from an assistive technology, boys with DMD would like to receive feedback about both quality and quantity of their movements. However, concerns were raised by the DMD group that the system needed regular reassessment in order to compensate for the deteriorating nature of the disease and the avoidance of overuse and pain.

In order to improve, people with stroke would like to be motivated by the system. In stroke, people can usually use their unaffected arm and hand unobtrusively to perform complex movements [31], which demotivates them to use their affected hand. Awareness of their movements and (possible) non-use of the affected side during daily life activities is important to them.

All patients, family caregivers and healthcare professionals were positive regarding the potential of assistive technologies to facilitate self-management and independence. Although the present study highlighted many similarities in essential topics to be covered in the design of an intelligent UL supporting assistive technology between DMD and stroke, interpretation of some of the specific requirements involved was different due to the differences in the target populations. Although differences in focus of the user perspectives may also be due to slight differences in information collection and participants between both groups, this study provides valuable information about users' views regarding technology aspects of an assistive technology, and relevant insights into the most population-specific topics. In this study, healthcare professionals were included in the prioritization of the user requirements afterwards, but they did not take part in the focus groups. Therefore, our findings as presented in this paper may be different if healthcare professionals were included earlier on in the process.

V. CONCLUSION AND FUTURE WORK

Actual use of a device can be related to the perceived usefulness and ease of use [32]. In order to improve the chance of acceptance, specific attention needs to be paid to ease of use and usefulness as well as a high priority should be given to accessibility and personalization of both hardware and software aspects of assistive technology [27]. The user requirements from this study, focused on technical domains, can be used to complement the existing information on user perspectives identified as important barriers and facilitators for UL assistive technology. End-user input from this study has highlighted differences in end-user preferences and needs between and within populations, which have to be taken into account into the future design of an UL supporting assistive technology. Currently, the identified user requirements are being taken into account during the design of an intelligent, adaptive, unobtrusive UL supporting assistive technology within the eNHANCE project, aimed at assisting people with DMD or stroke in independently performing UL daily life activities.

ACKNOWLEDGEMENTS

This project has received funding from the *European Union's Horizon 2020 research and innovation programme* under grant agreement No 644000. This result reflects only the author's view and the Commission is not responsible for any use of the information in this result.

REFERENCES

[1] S. Balasubramanian, J. Klein, and E. Burdet, "Robot-assisted rehabilitation of hand function," Current opinion in neurology, vol. 23, pp. 661-670, 2010.

[2] J. L. Herder, N. Vrijlandt, T. Antonides, M. Cloosterman, and P.L. Mastenbroek, "Principle and design of a mobile arm support for people with muscular weakness," Journal of rehabilitation research and development, pp: 591-604, 2006.

[3] P. N. Kooren et al.,"Design and pilot validation of A-gear: a novel wearable dynamic arm support," Journal of neuroengineering and rehabilitation, 2015.

[4] D. S. Nichols-Larson, P. C. Clark, A. Zeringue, A. Greenspan, and S. Blanton, "Factors influencing stroke survivors' Quality of life during subacute recovery," Stroke, pp. 1480-1484, 2005.

[5] A. G. Dunning and J. L. Herder, "A review of assistive devices for arm balancing," IEEE International Conference on Rehabilitation Robotics, pp. 1-6, 2013.

[6] P. Maciejasz, J. Eschweiler, K. Gerlach-Hahn, A. Jansen-Troy, and S. Leonhardt, "A survey on robotic devices for upper limb rehabilitation," Journal of NeuroEngineering and Rehabilitation vol. 11, 2014.

[7] L. A. Van der Heide et al., "An overview and categorization of dynamic arm supports for people with decreased arm function," Prosthetics and Orthotics International, pp. 287-302, 2013.

[8] J. Frappier, "Clinico-economic study of the JACO robotic arm for powered wheelchair users with upper-extremity disabilities," By Data4Actions inc. for Kinova, Canada, pp. 1-5, 2011.

[9] G. R. B. E. Romer, H. J. Stuyt, and A. Peters, "Cost-savings and economic benefits due to the assistive robotic manipulator (ARM)," Proceedings of the 2005 IEEE 9th International Conference on Rehabilitation Robotics, pp. 201-204, 2005.

[10] L. N. Gitlin, "Why older people accept or reject assistive technology," Generations, vol. 19, pp. 41-46, 1995.

[11] S. Demain et al., "Assistive technologies after stroke: self-management or fending for yourself? A focus group study," BMC Health Services Research, vol. 13, 2013.

[12] A. M. Hughes et al., "Translation of evidence-based assistive technologies into stroke rehabilitation: users' perceptions of the barriers and opportunities," BMC Health Services Research, 2014.

[13] R. Haworth, S. Dunscombe, and P. J. R. Nichols, "Mobile arm supports: an evaluation," Rheumatology, pp. 240-244, 1978.

[14] A. Kumar and M. F. Philips, "Use of powered mobile arm supports by people with neuromuscular conditions," Journal of rehabilitation research and development, pp. 61-70, 2013.

[15] B. Radder et al., "User-centred input for a wearable soft-robotic glove supporting hand function in daily life," IEEE International Conference on Rehabilitation Robotics (ICORR), pp. 502-507, 2015.

[16] C. Abras, D. Maloney-Krichmar, and J. Preece, "User-centered design," W. Encyclopedia of Human-Computer Interaction. Thousand Oaks: Sage Publications, vol. 37, pp. 445-456, 2004.

[17] C. Stehle and C. Albrecht-Buehler, "Developing more desirable products for stroke survivors," Topics in stroke rehabilitation, 15(2), pp. 109-117, 2008.

[18] J. He, et al., "RUPERT: a device for robotic upper extremity repetitive therapy," Engineering in Medicine and Biology Society, 27th Annual International Conference of the IEEE, pp. 6844-6847, 2005.

[19] S. Landsberger et al., "Mobile arm supports: history, application, and work in progress," Topics in Spinal Cord Injury Rehabilitation, pp. 74-94, 2005.

[20] Y. L. Yasuda, K. Bowman, and J. D. Hsu, "Mobile arm supports: criteria for successful use in muscle disease patients," Archives of physical medicine and rehabilitation, 67(4), pp. 253-256, 1986.

[21] E. C. Lu et al., "The development of an upper limb stroke rehabilitation robot: identification of clinical practices and design requirements through a survey of therapists," Disability & Rehabilitation: Assistive Technology, 6(5), pp. 420-431, 2011.

[22] Y. Shakya, and M. Johnson, "A mobile robot therapist for under-supervised training with robot/computer assisted motivating systems," Engineering in Medicine and Biology Society30th Annual International Conference of the IEEE, pp. 4511-4514, 2008.

[23] D. Benyon, P. Turner, and S. Turner, "Designing interactive systems: People, activities, contexts, technologies," Pearson Education, 2005.

[24] P. M. Boynton, T. Greenhalgh, "Selecting, designing, and developing your questionnaire," BMJ, pp. 1312–1315, 2004.

[25] P. M. Boynton, "Administering, analysing, and reporting your questionnaire," BMJ, pp. 1372–1375, 2005.

[26] D. Clegg, and R. Barker, "Case method fast-track: a RAD approach," Addison-Wesley ISBN 978-0201624328, 2004.

[27] G. B. Prange et al., "User requirements for assistance of the supporting hand in bimanual daily activities via a robotic glove for severely affected stroke patients," IEEE International Conference on Rehabilitation Robotics (ICORR), pp. 357-361, 2015.

[28] A. L. van Ommeren, G. B. Prange-Lasonder, J. S. Rietman, P. H. Veltink, and J. H. Buurke, "Preliminary extraction of themes from a review about user perspectives on assistive technology for the upper limb after stroke," Proceedings of the 4th International Conference on Neurorehabilitation, pp.323-327, 2016.

[29] A. I. Batavia and G. S. Hammer, "Toward the development of consumer-based criteria for the evaluation of assistive devices," Journal of Rehabilitation Research and Development, vol. 27, pp. 425-436, 1990.

[30] H. W. Mahncke, Bronstone, A., and Merzenich, M. M., "Brain plasticity and functional losses in the aged: scientific bases for a novel intervention," Progress in Brain Research, pp 81- 109, 2006.

[31] I. M. Bullock, J. Z. Zheng, S. De La Rosa, C. Guertler, and A. M. Dollar, "Grasp frequency and usage in daily household and machine and shop tasks," IEEE Transactions on Haptics, vol. 6, pp. 296-308, 2013.

[32] F. D. Davis, "User acceptance of information technology: system characteristics, user perceptions and behavioral impacts," International Journal of Man-Machine Studies, vol. 38, pp. 475-487, 1993.

Correlational Analyses among Personality Traits, Emotional Responses and Behavioral States Using Physiological Data from Wearable Sensors

Ruiying Cai, Ao Guo
Graduate School of Computer and Information Sciences
Hosei University
Tokyo 184-8584, Japan
{ruiying.cai.85, guo.ao.33}@stu.hosei.ac.jp

Jianhua Ma, Runhe Huang
Faculty of Computer and Information Sciences
Hosei University
Tokyo 184-8584, Japan
{jianhua, rhuang}@hosei.ac.jp

Abstract—**Mental health is crucial to the overall well-being of individuals, societies, and countries. In addition, personality traits, emotional responses and behavioral states caused by acute stress have a significant influence on mental health. However, correlations among personality traits, emotional responses, and behavioral states were analyzed only by manual reports in psychology. This might cause a problem in that data might not be objective. Therefore, the main purpose of this paper is to examine whether and how personality traits are associated with emotional responses and behavioral states using physiological data from wearable devices. In experiments, 38 male and female university graduates and undergraduates volunteered as participants, and each of them first completed a Big Five Inventory (BFI) questionnaire and then made a presentation in class, to get personality traits, emotional responses, and behavioral states, respectively. In the presentation, three wearable devices were used for emotional response data collection, including an Emotive Insight detecting electroencephalogram (EEG) data, a Spire Stone collecting respiration data and a Huawei Fit Watch to get the heart rate value. In detail, six attributes of emotional responses: focus, interest, relaxation, engagement, stress and excitement, and 8 attributes of behavioral states: smile, clench, blink, surprise, furrow, wink, breath, and heart rate were used to analyze the personality traits. As a result, correlational analyses have indicated associations among the personality traits, emotional responses, and behavioral states.**

Keywords-wearable sensors; personality traits; emotional responses; behavioral states; correlation.

I. INTRODUCTION

Mental health is crucial to the overall wellbeing of individuals, societies, and countries [1]. Body sensors can be applied to better monitor individual's psychological conditions. In addition, personality traits, which can be defined as habitual patterns of behavior and thought, have a significant influence on mental health. Correlation analysis between mental health and personality traits, using physical data from wearable sensors, is an interdisciplinary field spanning computer science and psychology.

In psychology, Penley and Joe [2] have proposed that there are strong associations among the personality traits, emotional responses and behavioral states. Unfortunately, their study was conducted only by analyzing the answers to specific questions, and thus might not be so objective.

In computer science, several studies have found that personality traits are related to many factors, such as text, nonverbal communication, social network, and physical behaviors. For example, in context detection, IBM has provided Personality Insights [3], a service that uses linguistic analytics to infer individuals' intrinsic personality traits, such as the Big Five model [4]. In addition, Gundogdu et al. [5] proposed that personality traits are related to social relationships by analyzing face-to-face interactions. What's more, personality traits are associated with physical activity, speech activity, and face-to-face time, was studied based on wearable sensors [6]. Besides these, with the development of sensors, such as EEG and Electrocardiogram (ECG), computing emotions become possible. More recently, ASCERTAIN [7] recognized personality traits by emotions using data collected by commercial sensors, like EEG, ECG, and Galvanic Skin Response (GSR), while participants were watching some affective movie clips. However, this work did not take a person's behaviors, such as smile and wink, into consideration.

Hence, this research aims to find out correlations among the personality traits, emotional responses and behavioral states using physiological data from wearable sensors. To the best of our knowledge, it is the first work to analyze associations among the three aspects of personality traits, emotional responses and behavioral states using wearable sensor data. A modified personalized mental health framework has been proposed to better monitor individuals' mental health, which has taken the personality traits into consideration.

The rest of this paper is organized as follows. Section II describes the personalized mental health framework. Section III explains data collection experimental details, which includes descriptions of participants, measurements of personality traits, and three wearable devices used for getting emotional responses and behavioral states. Section IV goes into correlational analyses among the personality traits, the emotional responses, and the behavioral states. The important findings of this study and necessary future work are depicted in the last section.

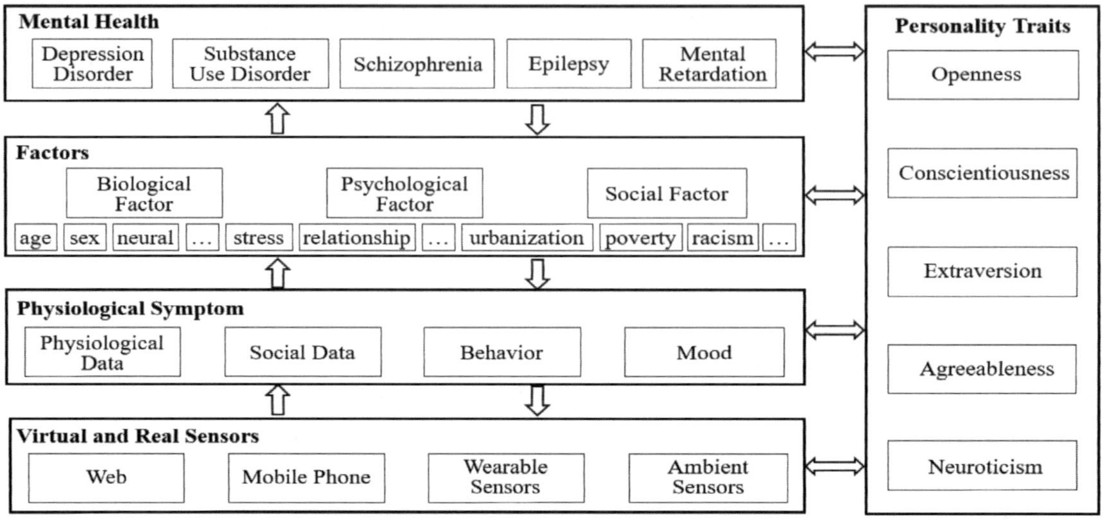

Figure 1. A personalized mental health framework.

II. PERSONALIZED MENTAL HEALTH FRAMEWORK

For better monitoring individuals' psychological conditions, a personalized mental health management framework was designed based on [8], shown in Fig. 1. In this personalized mental health framework, four layers are designed to transfer raw sensor data to mental health statements. In the first layer, raw data are collected from virtual or real sensors, for example, web browsing histories and accelerometers embedded in mobile phones. Then, the second layer aims to detect physiological symptoms, like the mood [9] and the behavior [8], based on these raw data. In the third layer, detailed factors are illustrated, which can be reported to professional doctors. Also, these data can be used to automatically detect psychological statements. On the other hand, this layer can be considered as a personal health log management layer. Finally, according to these factors, a mental health statement can be diagnosed by doctors or a mental health system. The key point of this framework is that personality traits were taken into consideration, which have an obvious influence on mental health. For example, an individual might get a stress notification sent by the traditional mental health system, because of a person's few number of activities. But, in case of an indoors loving person, with low openness value, maybe they enjoy time at home without many activities. In this situation, the traditional system will make a mistake because it did not take into account the personality traits. It is apparent that personality traits have associations with each layer, and this idea is assumed in this paper. Whether and how personality traits influence each layer is still ongoing study, not only in psychology but also in computer science. So, in this paper, we address these problems.

III. DATA COLLECTION

This section mainly explains data collection experimental details. It can be divided into four parts, participants to this study, an approach of measuring personality traits, three wearable devices, and attributes of emotional responses and behavioral states.

A. *Participants*

This research is composed of 38 university undergraduates and graduates. Each of them first completed a Big Five Inventory (BFI) questionnaire, and then took a presentation in class. When they were presenting, three wearable devices were equipped to record emotional responses and behavioral states.

B. *The Measure of Personality Traits*

The Big Five Model describes five personality traits: Openness (O), Conscientiousness (C), Extroversion (E), Agreeableness (A) and Neuroticism (N). In detail, personality traits are measured with a 44-item version of BFI questionnaire [10] and the range of O is [10, 50], C is [9, 45], E is [9, 45], A is [9, 45], and N is [8, 40].

C. *Wearable Devices*

Three devices are applied into this study. They are Emotive Insight, Spire Stone, and Huawei Fit watch.

1) Emotive Insight

Emotive Insight is a 5-channel mobile EEG headset that records a user's brainwaves and translates them into meaningful data. Six attributes of emotional responses, including focus, interest, relaxation, engagement, stress, and excitement, were recorded in real time. Each emotion contains 4 metrics: Min, Max, Raw and RawNorm values. In detail, Min, Max and Raw value are for scientific research and might be negative. RawNorm is scaled value deduced from the Raw, Min, and Max value to be bounded between 0 and 1. Besides emotional responses, this device also can recognize six attributes of behavioral states: smile, clench, blink, surprise, furrow and wink (left and right).

2) Spire Stone

Spire stone is a piece of equipment which supports continuous respiration sensing and real-time interventions. Respiration data can be considered as the seventh attribute of behavioral states.

3) Huawei Fit Watch

Although Huawei Fit is a smart piece of equipment, which can track sports activity, including sleep monitor, heart rate, and notification display, it is hard for developers to get real-time data. Thus, in this study, this device supports little. But for further study, the mobility of this device might give a hand. So, only heart rate can be considered as the eighth attributes of behavioral states.

D. Emotional Responses and Behavioral States

Due to associations among personality traits, emotional responses and behavioral states have a significant influence on mental health, it is very important to create a stressful scenario since it is strongly related with data quality. Therefore, a scenario of presentation in class was applied to this study according to [2].

Because of class time limitation, three classes of presentation were recorded respectively, and each class had 13, 13, and 12 student presentations, respectively. Each presentation was about 8 minutes and was composed of 5 minutes personal speech and 3 minutes of questions & answers. Besides personality traits, all emotional responses and behavioral states data were collected by wearable sensors.

Descriptions of six attributes of emotional responses are as follows.

- Interest measures on how much a person likes or dislikes something.
- Excitement captures the level of emotional arousal.
- Engagement measures how immersed a person is in what they are doing or experiencing.
- Focus is an ability to concentrate on one task and ignore distractions.
- Stress measures how comfortable a person is with the current challenge they are facing.
- Relaxation is an ability to switch off and reach a calm mental state.

As summarized, there are eight attributes of behavioral states that will be analyzed with personality traits.

- Smile: The possibility of smiling, ranged in [0, 1].
- Clench: The possibility of clench, ranged in [0, 1].
- Blink: The possibility of blink, ranged in [0, 1].
- Surprise: The possibility of surprised, ranged in [0, 1].
- Furrow: The possibility of a furrow, ranged in [0, 1].
- Wink left and wink right: Detecting wink activity, and the range is [0, 1].
- Breath: Number of breaths per minute, ranged in [10, 40], detected by Spire.
- Heart rate: Ranged in [30, 200], detected by Huawei Fit watch.

After data preprocessing, 90837 rows of data are extracted from raw and real-time data. On average, each participant contains 2390 rows of data, collected by Emotive Insight. Breath data are collected from the Spire server and heart rate data are got from Apple Health Kit.

IV. Correlational Analyses

The Pearson Correlation Coefficient (PCC), which measures the linear relationship between two variables X and Y, is applied into this study [11]. Pearson correlation coefficient, noted as r, varies between -1 and +1 with 0 implying no correlation. Positive correlations imply that as x increases, so does y. Negative correlations imply that as x increases, y decreases. Equation (1) shows, n is the sample size, x_i and y_i are the single samples indexed with i and $\bar{x}$, $\bar{y}$ are the mean value of x and y.

$$r = \frac{\sum_{i=1}^{n}(x_i - \bar{x})(y_i - \bar{y})}{\sqrt{\sum_{i=1}^{n}(x_i - \bar{x})^2}\sqrt{\sum_{i=1}^{n}(y_i - \bar{y})^2}}. \quad (1)$$

To test the null hypothesis that the true correlation coefficients equal to 0, based on the value of the sample correlation coefficient r, permutation tests provide a direct approach to performing hypothesis tests and constructing confidence intervals. Here, p-value is desired to test the null hypothesis and it is the proportion of the r values generated in permutation test that are larger than the PCC that was calculated from the original data. In detail, $p <= 0.01$ implies strong evidences to reject null hypothesis, $0.01 < p < =0.05$ implies evidences to reject null hypothesis, $0.05 < p <= 0.1$ implies some weak evidences to reject null hypothesis, and $p > 0.1$ means no evidence to reject null hypothesis. Here, null hypothesis is that there is no correlation between two parameters X and Y.

Thus, 38 students' personality trait values are shown in Fig. 2. Each person has 5 dimensions: Openness Conscientiousness, Extraversion, Agreeableness, and Neuroticism. In correlational analyses, mean values of emotional responses and heart rate, sum values of smile, clench, blink, surprise, furrow, wink left and right, and breath were analyzed with personality traits. Table I presents meaningful Pearson correlation coefficients between each dimension of Big Five and the mean values of six attributes emotional responses.

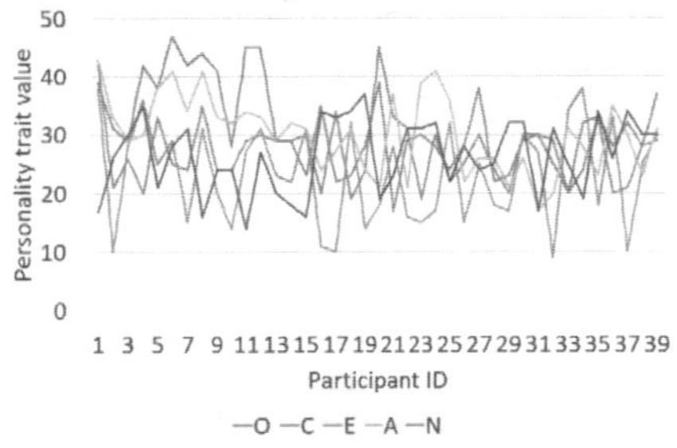

Figure 2. Personality values of participants.

TABLE I. CORRELATIONS BETWEEN THE BIG FIVE AND EMOTIONAL RESPONSES

Emotional Responses	Big Five Dimensions				
	O	*C*	*E*	*A*	*N*
EngagementRaw	0.07	-0.04	0.08	0.20	**0.43****
FocusRaw	0.13	-0.03	-0.04	0.07	**0.49***
InterestMax	-0.09	**0.37***	-0.07	**-0.49****	**-0.56****
InterestMin	0.13	**-0.37***	0.02	**0.42****	**0.56****
InterestRaw	-0.13	**0.34***	-0.02	-0.24	-0.25
ShortTermExcitementRaw	0.13	-0.03	-0.04	0.07	**0.49***
StressMin	-0.29	0.17	0.04	-0.02	**-0.32***

Note. *p<0.05, **p <0.01

Table II presents meaningful PCC between each dimension of Big Five and the sum value of smile, clench, blink, surprise, furrow, wink left and right, breath and heart rate. Some meaningful results will be shown as follows.

- Openness (O): Openness was positively associated with wink.
- Conscientiousness (C): Conscientiousness is most associated with interest. In addition, the smaller sum value of smile and wink, the more conscientious.
- Extraversion (E): Extraversion has little influence on emotional responses and behaviors.
- Agreeableness (A): The smaller range of interest value, the more agreeable. Besides, agreeableness is most relevant to behaviors. The greater sum value of blink, wink, and breath, the more agreeable. In addition, agreeableness is negatively relevant with clench and heart rate.
- Neuroticism (N): N is most relevant to emotional responses. The easier engaged, focused, interested, and excited, the more neurotic. In addition, neuroticism was negatively correlated with stress. For behaviors, the more blink and greater heart rate, the more neurotic.

TABLE II. CORRELATIONS BETWEEN THE BIG FIVE AND BEHAVIORS

Behavioral States	Big Five Dimensions				
	O	*C*	*E*	*A*	*N*
Smile	0.23	**-0.34***	0.15	-0.06	-0.0
Clench	-0.10	-0.12	-0.05	**-0.34***	-0.24
Blink	0.10	-0.29	0.22	**0.34***	**0.40***
Wink Left	**0.35***	**-0.36***	0.25	**0.37***	0.27
Wink Right	0.07	-0.23	-0.08	**0.33***	0.29
Breath	-0.21	0.26	0.29	**0.60***	0.08
Heart Rate	-0.10	-0.08	0.11	**-0.53***	**0.49***

Note. *p<0.05, **p <0.01

V. CONCLUSIONS AND FUTURE WORK

To summarize, our correlational analyses in this study revealed two main points. The first one is that Neuroticism (N) is most related to emotional responses and this conclusion is also supported by previous psychological researches. Secondly, agreeableness is most related to behaviors. But there are still some problems in this work, such as unstable connection of the Emotive Insight. In detail, sometimes participant's movement or hair might not guarantee a high quality of connection. Based on these two conclusions, further work is needed on how personality traits are associated with emotional responses and behavioral states.

ACKNOWLEDGMENT

This work is partially supported by the Japan Society for the Promotion of Science Grants-in-Aid for Scientific Research (No.26330350).

REFERENCES

[1] J. Sayers, "The world health report 2001 — Mental health: new understanding, new hope," Bulletin of the World Health Organization, vol. 1085, pp. 5-16, Nov. 2001.

[2] A. J. Penley and T. Joe, "Associations among the Big Five, emotional responses, and coping with acute stress," Personality and Individual Differences, vol. 32, no. 7, pp. 1215-1228, 2002.

[3] J. Mahmud, "IBM Watson personality insights: The science behind the service," Technical report, IBM, 2016.

[4] P. O. John and S. Sanjay, "The Big Five trait taxonomy: History, measurement, and theoretical perspectives," Handbook of Personality: Theory and Research, vol. 2, pp. 102-138, 1999.

[5] D. Gundogdu, N. A. Finnerty, and J. Staiano, S. Teso, A. Passerini, F. Pianesi, and B. Lepri, "Investigating the association between social interactions and personality states dynamics," Royal Society Open Science, vol. 1, pp. 1-3, Oct. 2017.

[6] O. D. Olguin, P. A. Gloor, and P. Alex, "Wearable sensors for pervasive healthcare management," Pervasive Computing Technologies for Healthcare, 3rd International Conference on IEEE, vol. 1, pp. 1-4, Apr. 2009.

[7] R. Subramanian, J. Wache, M. Abadi, R. Vieriu, S. Winkler, and N. Sebe, "ASCERTAIN: Emotion and personality recognition using commercial sensors," IEEE Transactions on Affective Computing, 2016.

[8] H. W. Lee and J. K. Chun, "The relationship between depression, perceived stress, fatigue and anger in clinical nurses," Journal of Korean Academy of Nursing, vol. 36, no. 6, pp. 925-932, 2006.

[9] C. D. Mohr, Z. Mi, and M. S. Stephen, "Personal sensing: Understanding mental health using ubiquitous sensors and machine learning," Annual Review of Clinical Psychology, vol. 13, pp. 23-47, May. 2017.

[10] D. F. B. Facet, "Big five inventory (BFI)," Differences, vol. 54, pp. 4-45, 2000.

[11] I. Lawrence and K. Lin, "A concordance correlation coefficient to evaluate reproducibility," Biometrics, pp. 255-268, 1989.

Worklife Ergonomics in eHealth Co-Creation Governance

"You can't manage what you don't measure"

Niels F. Garmann-Johnsen[1,2], Migle Helmersen[1]
[1]Agderforskning
Kristiansand, Norway
Emails: {niels.frederik, migle.helmersen}@agderforskning.no

Tom Roar Eikebrokk[2]
[2]Dept. of Information Systems, University of Agder
Kristiansand, Norway
Email: tom.eikebrokk@uia.no

Abstract—**This is a conceptual article where we seek to combine the concepts of information systems frameworks and newer co-creation literature as a means to devise a model for servitization and digitalization. Servitization and digitalization are two megatrends that affect healthcare and public services along with other sectors in general. A new model is needed to prescribe how governance in an increasingly changing world of modern healthcare could be undertaken in a successful manner. The concept of good worklife ergonomics is studied, both as a prerequisite and as a success factor. This article proposes that the moderating, risk mitigating, factor of broadly based employee involvement in all phases from planning and design to implementation will greatly improve quality in both innovation-process, and outcomes. A case-study from a public homecare living lab eHealth-project in Norway is visited to highlight some of the challenges ahead. Our discussion and conclusion end up devising the proposed model, and further research into how this model can be implemented is recommended.**

Keywords-Co-creation; servitization; digitalizaton; healthcare; eHealth; worklife ergonomics.

I. INTRODUCTION

The purpose of this article is to develop a conceptual process-model for co-creation in eHealth innovation, that also supports a good worklife ergonomics for employees. The article is a result of a cross-disciplinary collaboration, between one medical doctor, specializing in health and work environment, and two doctors of philosophy in social science, with management information systems as speciality.

In many industrial countries, people live longer, but habitually with chronic diseases, due to better living standards and medical treatment advances. These changing population demographics mean there is an increasing demand for healthcare services [1]. eHealth technologies accompanied by changes in healthcare delivery processes and services, offer possibilities for a lower cost healthcare system, needed to meet future increases in demand for services. These changes can be referred to as servitization transformation [2] and put an emphasis on the interaction with customers that requires providers to offer customized and total solutions [2]. Digitalization capabilities support such servitization through employees' involvement and codetermination of what should count as key performance indicators. Digitalization is "the use of digital technologies to change a business model and provide new revenue and value-producing opportunities; it is the process of moving to a digital business" [3].

But, such change-projects often meet unforeseen barriers. Objections may be raised by the various professional groups themselves. Poor rooted changes risks leading to inferior solutions over time, which may work against their purpose. In Norway, primary care and homecare is a concern for the public sector; municipalities. Generally, new technologies and working methods, as well as new service providers, will have to absorb all the "tacit knowledge" inherent in the public organization to add new values to the services in an efficient manner. If employees are involved, they may be more inclined to become a driving force in the pursuit of a servitization strategy, that relies on developing digitalization capabilities, because the process of defining performance criteria promotes organizational learning [4].

In Norway, primary healthcare and homecare is the concern of the municipalities. Local government-initiated eHealth pilot-projects are often unconnected experiments. A shared and common process management methodology for both development and implementation phases, that incorporates employee involvement and collaboration, will arguably be a useful tool for public sector change leaders who want to introduce new technologies and working methods, or invite new service providers that relieve or complete the overall welfare offer to citizens. We will term this as Co-creation governance ('Co-creation' as a term is disseminated further in Section IV). Such a tool will be useful in the complex task of maintaining quality for both service recipients and service providing personnel – employees in the healthcare system. Lenka et al. [2] has recently proposed a model for co-creation between a product or service vendor and end-consumers. But, in eHealth innovation in the Norwegian context, system vendors, and health care providers are most often separate entities, so the health care provider generally add value through the combination of human services and the application of technology, not technology alone. Seen from the point of view of the health care provider, the research and development challenge can be put as:

- How employees' involvement is ensured in eHealth co-creation governance?
- How this involvement contributes to ensuring performance quality on all levels of responsibility?

A potential solution to this challenge is the development of a shared digital capability to continually improve service quality. When in place, this capability will ensure that internal and external service producers act through a continuous quality improvement cycle from plan, check, act, and correct that improves service quality over time. This understanding of digital capabilities is in line with Lenka et al.'s model. But this article argues that this capability must be developed along two dimensions of co-creation or collaborative innovation:

- Horizontally along the chain of value co-creation, from ICT-vendor, through service-provider to home care service users, but also
- Vertically along a line of innovation-process governance, from front-stage service-personnel employees to top-management.

These capabilities must subsequently be built "bottom up" with the involvement and participation of all relevant municipal employees, ensuring that new and increasingly more technology enabled work processes still remain employee friendly, and thus improve the quality of worklife of employees as well as patients' quality of life [5]. This article will also show how Lenka et al.'s aspects of digitilization, servitization and co-creation are linked to our highlighted aspects of (worklife) ergonomics, Business Performance Management, and (Information System) governance.

Ergonomics is an applied science concerned with designing and arranging things people use so that the people and things interact most efficiently [6]. Ergonomics is the science of designing the workplace, keeping in mind the capabilities and limitations of the worker and in such way, fulfil the goals of occupational health and safety, and productivity of employees [7]. The implementation of new digital services in healthcare involves several new work tasks, and thus represents new work processes and potential risk factors at the workplace. Knowledge of this should be addressed to prevent potential negative health effects among employees. This article proposes the term worklife ergonomics as a holistic term that encompasses the system of service production that spans over workplaces and involved employees. As such, worklife ergonomics as a concept considers the whole information system with people, processes and technology. Employee engagement and involvement brings a new and needed perspective into co-creation servitization, and digitalization.

Effective Business Performance Management, and (Information System) governance are important factors in achieving successful innovation, and the authors will show that such management tools need to be activated in parallel with the system- and service development processes. Employee involvement in the creation and execution of such management tools will serve to ensure the goals are met, and risks for failure are mitigated.

The rest of this article is structured as follows: In Section II, the authors shortly renders the methods used; literature search, and a comparison of the findings from this, with a research project performed in a Living Lab context. In Section III, the results from the literature search is presented; ergonomics in organizational change in general (A), and in the context of eHealth (B). In Subsection C, the result of the comparisons with application of concepts the case study is presented. In Section IV, using the Lenka et al. model as a guide, we propose a model for governance that incorporates principles for good worklife ergonomics in eHealth. In Section V, we conclude that the proposed model may contribute to meeting the research challenges proposed in this introduction and also devise avenues for further research.

II. Method

To devise a conceptual model of worklife ergonomics, we conducted a literature review to explore how ergonomics are used in relation to the concepts of eHealth, digitalization and co-creation. The authors were looking for principles in the literature that could guide us conceptually in designing a system that would encompass good worklife ergonomics.

A literature review was performed in October 2017. Using Google scholar, the literature was searched for articles containing the criteria (search string); ergonomics AND digitalization AND servitization AND health AND employees. By using such Boolean-logic operators; 'AND', the authors ensured that the findings where narrowed to only articles including all the key-terms, thus covering the desired context. This search and screening, resulted in three articles that provided concepts with substantially new insight (the rest of the articles screened, only briefly touched the key criteria).

The identified concepts from literature (see III, Results, Section A and B), were compared with findings from discussions from awareness-workshops in an eHealth Living Lab action research project in a municipality in Norway (see III, Results, Section C). In this project, the research team (including the authors of this article) held six awareness workshops together with representatives from the municipality (a joint project manager, ICT manager, management and employee representatives from municipal homecare and nursing services). The workshops focused on these topics:

- Stakeholder analysis
- Service design and 'design thinking' methods
- ICT-business as innovation partners (ref. co-creation with ICT-system vendors)
- Capabilities and organizational learning
- Enterprise performance management, and
- Scaling up innovations from a Living lab.

Two of the authors also visited design workshops where front-line personnel employees in home nursing, together with municipal healthcare-department managers and eHealth researchers, discussed issues and requirements related to a specific service innovation, the use of digital night surveillance for patients in need of this, staying at home, with use of cameras with video conferencing functionalities.

III. RESULTS

The results are presented in relation to the key terms of the literature search. The identified articles offered design principles that can govern good worklife ergonomics in eHealth co-creation processes.

A. Ergonomics in co-creation – the role of employees

Neubauer and Stary [8] describe ergonomics as acknowledging the role of employees in innovation as leading to both improvements and financial benefits, through human-centred design.

Human-centred design for interactive systems promotes the following key principles [8]:

- The design is based upon an explicit understanding of users, tasks and environments
- Users are involved throughout design and development
- The design is driven and refined by user-centred evaluation
- The process is iterative
- The design addresses the whole user experience
- The design team includes multidisciplinary skills and perspectives.

Of advocation policies that could improve on this, Lopez-Gomez et al [9] suggest:

- Promoting the access to highly qualified personnel to develop new concepts and service innovations in-house
- Developing training methods for personnel to be able to adapt innovations acquired from external sources
- Need to better adapt curricula in education and training schemes to the demands of service economy
- Recognizing informal learning so as to increase the attractiveness of continuous training for employees
- Promoting modern innovation management approaches that better support creativity and autonomy of service workers [9]

B. Operationalizing these principles in eHealth

While the forgone citations are from industrial contexts, Beaumont et al [1] focus on service-design in eHealth, and propose that socio-technical, human-centred design approaches are better alternatives to techno-centric design. The article promotes joint innovation tools like service blueprints [10][11] and stakeholder analysis [5] in the form of Systems Scenario Tool (SST) [12]. SST combines stakeholder, and system gap-analysis.

The key points in the article are [1]:

- Telehealth equipment and services offer opportunities for bridging the future gap between available health resources and demand created by an increase in life expectancy.
- Current use of telehealth is limited by inadequate business models and service designs that fail to generate successful partnerships and value for customers and suppliers.
- Traditionally, healthcare providers have taken a techno-centric approach to the implementation of new technologies, which often results in unforeseen barriers to success.
- Design and implementation of new services can benefit from a socio-technical approach, which gives equal consideration to both social and technical aspects of a complex system.
- Co-creation of value requires new tools, such as the System Scenarios Tool, which provides stakeholders with a holistic framework to help model the implications of service offering and business model choices.

C. Design principles applied on a Living Lab project

Comparing these organizational design principles with experiences from the Living Lab project workshops, methods such as stakeholder analysis [5], and service blueprints (op. cit.), as devised in Beaumont et al, found in the literature review [1], proved to be useful in helping to design new services. To the known service blueprint template for process notation (swim lane diagram) we found it useful to add a band for step purpose and key performance indicators, see Fig. 1. In addition to showing the process following a timeline or sequence (steps), the process diagram shows activities at different levels of the information system. The levels include both those parts that are visible to the end-user and the processes back stage, below the "line of visibility" [11]. Adding the purpose of each step purpose makes it possible to extract user stories to form a system requirement documentation for hand-over to Information Technology Infrastructure Library methods (ITIL) [13]-[15] or agile system development [16], and refined further to precise technical architectures and instructions to ICT-system engineers. At the same time, adding key performance indicators can be a starting point for defining inputs to a joint enterprise process and performance management system.

Service Blueprint diagram	Pre service	During service			Post service
Step name/no.					
Step purpose					
Key performance indicators					
Service evidence					
User action					
Front stage personnel action	↓ *"Visibilityline"*				
Back stage personnel action					
Infrastructures (legal, standards, technical)					
Legend	Event	Process	Descision	Connection	

Figure 1. Service Blueprint diagram template, with "lanes" for purpose and performance indicators added

By adding the iteration of a workshop with all involved front-stage and back-stage personnel-employees, like in the Living lab-case (see Section II; Method), more aspects of a proposed innovation can be explored, before expensive investments and changes are made. Although our process modelling exercise showed the proposed camera-surveillance case to be technically feasible and may give potential benefits to homecare patients, it also showed that such an innovation also has major implications for the worklife for e.g., home nurses, as well as legal and privacy-issues in general, that needs to be examined and discussed further. The status, as this article is being written, is that the camera-surveillance case has been postponed, while other innovation-paths are explored; e.g., a new contact-centre and alarm-reception central.

IV. Proposition

This article proposes that enterprises that want to succeed with eHealth innovation and co-creation over time, need to secure the involvement of their frontline personnel, because they are key to establishing a Business Performance Measurement system. There are numerous definitions of what a Business Performance Measurement system contains. In a literature review, Franco-Santos et al. [17] identified these main features [17]:

1. Performance measures
2. Objectives/goals
3. Supporting infrastructures (including data acquisition and analysis)
4. Targets (gauges – does the enterprise meet its targets)
5. Causal models (what are drivers for successful performance)
6. Hierarchy/cascade (organization, delegation of concern)
7. Performance contract (negotiated contractual relationships with stakeholders)
8. Rewards (incentives)

Co-creation is a relatively new term. It has become part of the slogan and strategy of many universities. But what does it mean in practice and where does the term come from? A recent review by Galvagno and Dalli [18] traces the term back to three theoretical perspectives including service science, innovation and technology management, and marketing and consumer research. The literature on co-creation operates on two levels of analyses: company centred vs. customer experience centred. Apparent themes in the literature include co-creating value through customer experience and competence, service innovation, including digital customer involvement. Today, service science and marketing play a major role in the literature and refer to the involvement of customers in the supplier's product- and service development. In information systems research and management research, the term co-creation has been used by, among others, Grönroos and Voima [19], and Lenka et al. [2].

Lenka et al. have provided a model that will explain the connection between "megatrends" in industry and working life; digital development and change ("digitalization") and development of a service culture in production-oriented environments ("servitization") through co-creation processes. As authors, we agree with Lenka et al., that an important prerequisite for success is the development of digitalization capabilities in service-based organizations. These digitalization capabilities in turn, will govern the "Value Co-creation" mechanisms; consisting of two main mechanisms; one linked to needs analysis (perceptive mechanisms) and one linked to design and construction cycles (responsive mechanisms). Between these two (from observation to design and construction), knowledge about measurement points is transferred to goals and values that form the basis for implementation of the service (in design and construction). Both mechanisms must be repeated for each overlapping link in the value chain.

Moreover, we propose that the change work done in these overlapping links in the value chain can be expressed (including the core, the actual digitalization capability) as Deming Cycles (Plan-Do-Study-Act), see Fig. 2.

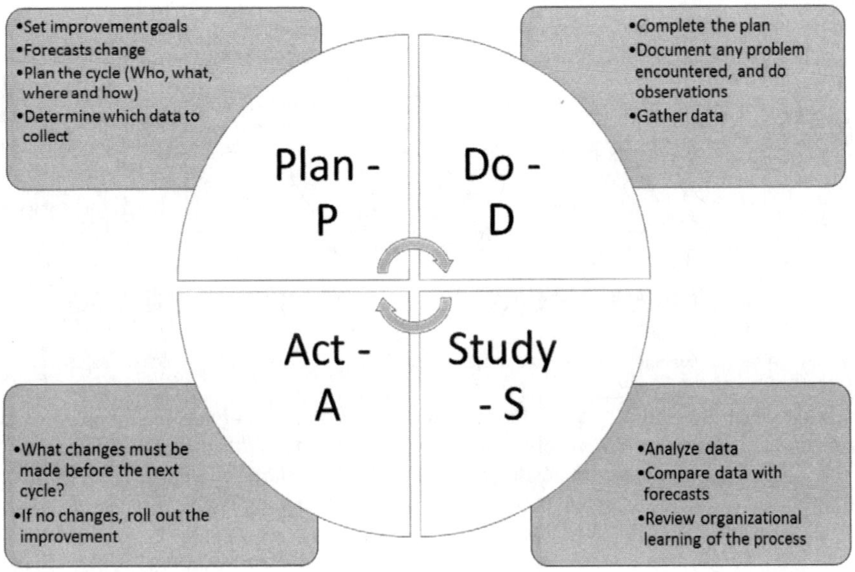

Figure 2. Deming cycles

Iterative development-cycles like this allow the time for involvement of both external and internal stakeholder groups, and should include discussing goals and measurements. The saying "You can't manage what you don't measure", referring to our sub-title, is attributed to both W. Edwards Deming and Peter Drucker.

Focus in achieving worklife ergonomics will be the relationship between the observational input and response outputs from the service co-creation and system co-creation cycles as a prerequisite for successful eHealth co-creation governance. Lenka et al. states, that value is added at each part of the chain, as new actors bring in new experiences, see new opportunities and add new value to the service. This includes the service consumers themselves, and their next-of-kin. The measurement system will be a trigger for new innovations, while being a missed "GPS" guidance system, to find the way [20]. Such a system will also act to spur organizational learning, providing incentives that motivate and intensify innovation [20].

To stay relevant, since the frames, and context, of the eHealth area is rapidly changing, we believe that the overall quality system (Process and performance management system) itself must be agile and subject to at least annual evaluation (a slower Deming cycle), while the services that the system controls, go through its many and fast Deming cycles. Together, these form a proposed conceptual process-model for co-creation in eHealth innovation, that also supports good worklife ergonomics. The concept is illustrated in Fig. 3.

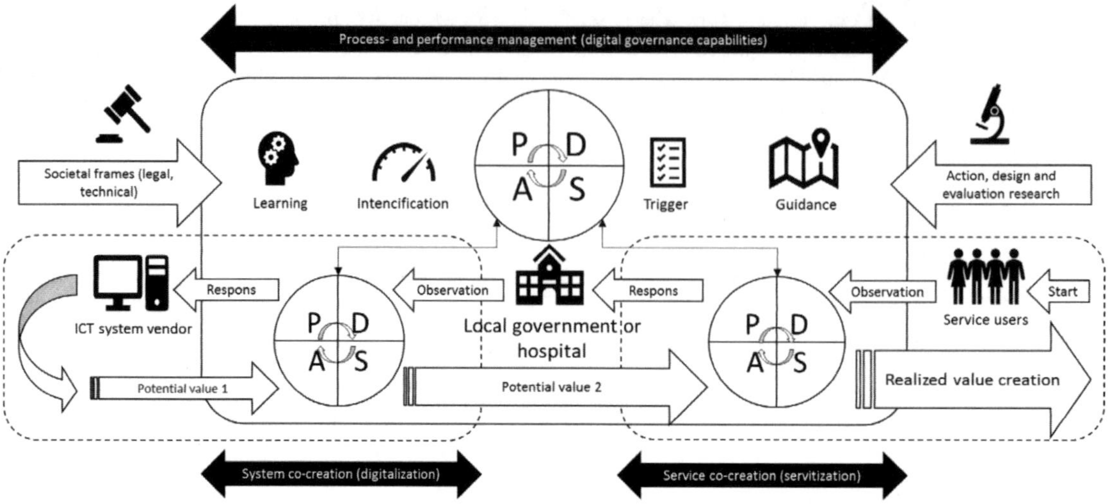

Figure 3. Process-model for co-creation in e-Health innovation; The ecosystem (Based on Lenka et al., 2017)

When it comes to the Plan-Do-Study-Act cycles in the value-chain (system co-creation, service co-creation) – different process modelling tools, like swim lane diagrams, can be used to visualize, convey, and discuss the consumer journey with stakeholders, using e.g., "Service Blueprint" or similar [10]-[21].

All in all, the goal is that the entire ecosystem is set in a state of continuous improvement and value innovation, and that a shared and improved service culture in the municipalities and their partners, (servitization), develops through digital transformation. This secures the ability to go back and start again, if necessary.

V. CONCLUSION AND FUTURE WORK

This article has shown that for the purpose of establishing a Living Lab concept, and achieving digitalization and servitization throughout the organization, the following elements are necessary; digital governance capabilities, process and performance management methods and systems that align new technologies with high quality work processes (worklife ergonomics), and appropriate tools to visualize and communicate processes and services with end-users, as well as different professional employee groups involved, front stage and back stage. By involving employees though iterative project-cycles and achieving a general consensus on what goals and measures should count, the necessary sorting and maturing of ideas is achieved, so that failed changes can be avoided before too great investments are made and lost.

Other factors that are necessary are processes that align local service strategies with central government legal and technical frames (compliance). In Norway, much of this is still in the making, and health information systems are generally not interoperable, but a joint information infrastructure is under development in mid-Norway (a joint health information platform and the "One resident – One joint health journal" project), and is expected to go national in 2022 [22]. More research is needed on how these different eco-systems (central, local) can be efficiently combined.

More research is needed into innovative means of capturing both qualitative and quantitative data about end-users or patients' using "Big data"; combining e.g., social media and transaction data from the service systems. More action, design and evaluation research are also needed for devising how the proposed model (Fig. 3) can be implemented and operationalized in a manner that ensures both employee and end-user involvement and commitment for achieving a high quality, lower cost health care system, while maintaining a high quality of worklife.

REFERENCES

[1] L. Beaumont, L. E. Bolton, A. McKay, and H. P. N. Hughes, "Rethinking service design: a socio-technical approach to the development of business models", in Product Development in the Socio-sphere2014, Springer. pp. 121-141.

[2] S. Lenka, V. Parida, and J. Wincent, "Digitalization Capabilities as Enablers of Value Co - Creation in Servitizing Firms". Psychology & Marketing, 2017, vol. 34, no. 1, pp. 92-100.

[3] Gartner, IT glossary. 2017 [cited 2018 Jan.]; Available from: https://www.gartner.com/it-glossary/digitalization/.

[4] R. G. F. Braz, L .F. Scavarda, and R. A. Martins, "Reviewing and improving performance measurement systems: An action research," International Journal of Production Economics, 2011, vol. 133, no. 2: pp. 751-760.

[5] N. F. Garmann-Johnsen and Ø. Hellang, "Collaborative Process Modelling and Evaluation in E-health". in Scandinavian Conference on Health Informatics; August 22;

2014; Grimstad; Norway. 2014. Linköping University Electronic Press.

[6] Merriam-Webster, Dictionary. 2017 [cited 2018 Jan.]; Available from: https://www.merriam-webster.com/dictionary/ergonomics'.

[7] L. Punnett et al., "A conceptual framework for integrating workplace health promotion and occupational ergonomics programs." Public Health Reports, 2009, vol. 124, no. 4 (suppl. 1): pp. 16-25.

[8] M. Neubauer and C. Stary, S-BPM in the Production Industry. 2017: Springer.

[9] C. López-Gómez et al., "Emerging trends in global advanced manufacturing," 2013, University of Cambridge: Policy links.

[10] G. Lynn Shostack, "How to design a service." European Journal of Marketing, 1982, vol. 16, no. 1, pp. 49-63.

[11] M. J. Bitner, A. L. Ostrom, and F. N. Morgan, "Service blueprinting: a practical technique for service innovation." California management review, 2008, vol. 50, no. 3, pp. 66-94.

[12] H. P. Hughes, C. W. Clegg, L. E. Bolton and L. C. Machon, "Systems scenarios: a tool for facilitating the socio-technical design of work systems". Ergonomics, 2017: pp. 1-17.

[13] T. R. Eikebrokk and J. Iden, "ITIL implementation: The role of ITIL software and project quality." in Database and Expert Systems Applications (DEXA), 2012 23rd International Workshop on. 2012. IEEE.

[14] J. Iden and T. R. Eikebrokk, "Using the ITIL process reference model for realizing IT Governance: An empirical investigation." Information Systems Management, 2014, vol. 31, no. 1, pp. 37-58.

[15] J. Iden and T. R. Eikebrokk, "Implementing IT Service Management: A systematic literature review." International Journal of Information Management, 2013, vol. 33, no. 3. pp. 512-523.

[16] R. Hoda, J. Noble, and S. Marshall, "Self-organizing roles on agile software development teams." IEEE Transactions on Software Engineering, 2013, vol. 39, no. 3, pp. 422-444.

[17] M. Franco-Santos et al., "Towards a definition of a business performance measurement system" International Journal of Operations & Production Management, 2007, vol. 27, no. 8, pp. 784-801.

[18] M. Galvagno and D. Dalli, "Theory of value co-creation: a systematic literature review." Managing Service Quality, 2014, vol. 24, no. 6, pp. 643-683.

[19] C. Grönroos and P. Voima, "Critical service logic: making sense of value creation and co-creation." Journal of the academy of marketing science, 2013, vol. 41, no. 2, pp. 133-150.

[20] A. Pavlov and M. Bourne, "Explaining the effects of performance measurement on performance: An organizational routines perspective." International Journal of Operations & Production Management, 2011, vol 31, no. 1, pp. 101-122.

[21] J. Schneider and M. Stickdorn, "This is service design thinking: basics, tools, cases" 2011: Wiley.

[22] Helse-Midt-Norge, "Én journal for hele helsetjenesten i Midt-Norge (in English: One journal for the entire health service in central Norway)." 2017 [cited 2018 Jan.]; Available from: https://helse-midt.no/vart-oppdrag/prosjekter/ehelse/helseplattformen#slik-er-helseplattformen-organisert.

OPeN:

Linking the National Adverse Reactions Database with Clinical IT Systems in Croatia

Arian Rajh, Dubravka Sudic, Katarina Gvozdanovic
Agency for Medicinal Products and Medical Devices
Zagreb, Croatia
e-mail: {arajh, dsudic, kgvozdanovic}@halmed.hr

***Abstract*— The Croatian Agency for Medicinal Products and Medical Devices is in the middle of implementation of project "OPeN." With its completion, the Agency will automate capturing of pharmacovigilance data from various clinical Information Technology (IT) systems and enable data syndication in the Croatian National Adverse Reactions database. The mechanism can consequently help healthcare professionals to avoid repeated input of data and save their time; this way, it will enhance medical practice and improve the public health system in Croatia.**

Keywords*— *Adverse Drug Reactions; Clinical Information Technology Systems; National Adverse Reactions Database; OPeN; Pharmacovigilance.

I. INTRODUCTION

The Agency for Medicinal Products and Medical Devices (HALMED) [1] is the National Competent Authority (NCA) in charge of the regulation of medicinal products and medical devices in Croatia. Its post-approval activities include monitoring of pharmacovigilance. "Pharmacovigilance is the science and activities relating to the detection, assessment, understanding and prevention of adverse effects or any other medicine-related problem." [2] An Adverse Drug Reaction (ADR) is a noxious and unintended drug effect [3]. HALMED is responsible for the maintenance of the national ADR database.

Pharmacovigilance (PhV) is principally a data-driven discipline. The database contains all ADRs received from various sources in Croatia. There were many advances in the last decade related to pharmacovigilance in general and ADR reporting in a particular degree.

Although spontaneous reporting is a "valuable resource for detecting actual drug-drug interactions" [4], the global PhV community has recognised the problem of underreporting [5]. That led to taking regulatory actions, as reported in European Medicines Agency's publications [6]–[8], as well as international guidance [9], further prescribed by international specifications [10] [11]. It also led to the use of technological innovations like data mining [12] [13] or other IT techniques, and finally to conducting standardisation efforts in the field [14]–[16]. One way to mitigate the problem of underreporting is to get, or to mine, large quantities of data [17]. Another and additional way to do it is to add relevant [18] input from health professionals into the reporting process and to increase the quality of data.

To enhance its pharmacovigilance capacities, HALMED participated in international pharmacovigilance IT projects – "Strengthening Collaboration for Operating Pharmacovigilance in Europe Joint Action" (SCOPE) and "WEB-Recognising Adverse Drug Reactions" (WEB-RADR). The SCOPE project, a joint action at European Union (EU) level, focused on the development of reporting mechanisms for ADRs [19]. It resulted in several IT-related recommendations related to facilitation of data collection/syndication and processing in various systems. Croatia, Netherlands, and United Kingdom developed mobile apps for reporting ADRs under WEB-RADR in 2015 and 2016 [20].

HALMED gained experience through these international projects and initiated a new project called "OPeN." [24] OPeN conformed entirely to the third suggestion of *Work Package 4* of project SCOPE, namely its recommendation to NCAs to "integrate suspected ADR reporting into clinical IT systems" [21].

The goal of the OPeN project was to increase both the number and the quality of ADR reports and to facilitate the communication between Health Care Professionals (HCPs) and NCAs via connected clinical IT systems and national ADR databases.

This paper describes the development of HALMED's OPeN system. In Section II, we describe risks related to the OPeN project, as well as project phases and methods of system development. In the same section, we describe modules of the system and its use. In Section III, we discuss why and how certain pharmacovigilance mechanisms were included in efforts taken by HALMED during the past decade, as well as in the OPeN project itself. We explain the advantages of direct reporting and the gist behind HALMED's project. We also outline examples of equivalent European projects, selected for study because they use mechanisms of direct ADR reporting from clinical systems. Despite the fact that they were examined before the development of HALMED's own system, we would nevertheless like to highlight some features like integration or communication with other systems, as well as coding and data mapping abilities. In Section IV, we show expected benefits of the OPeN project for Croatian public health system. Finally, Section V announces future work on the project, adding a goal to further facilitate the Croatian health sector.

II. Methods of the Development

A. *Project risks, phases, and the development*

Establishing a system as complex as OPeN involves many different institutions and stakeholders (The Ministry of Health of the Republic of Croatia, The Croatian Health Insurance Fund, The Croatian National Institute of Public Health, and providers of IT solutions and services in the healthcare domain). The most significant risk in this entire process is ensuring the necessary funding, as well as coordinating activities of various stakeholders, on time. Within the extent of the OPeN project, it was planned to connect OPeN with the Central Health Information System of the Republic of Croatia (CEZIH) [22], as well as with different hospital IT systems. However, this was not yet accomplished, and HALMED's staff is actively working to achieve it. A possible risk related to this integration attempt could be the circumstance that the operation of CEZIH is not within the remits of HALMED (the responsibility lies with The Croatian Health Insurance Fund). Another risk is related to the fact that there are 64 hospitals in Croatia: they have various levels of computerisation and use various IT solutions. HALMED has developed the OPeN system to the extent that it is functionally ready; however, other Croatian main stakeholders need to incorporate it into the national healthcare ecosystem. The ideal path would be to enforce providers to incorporate ADR reporting elements into their clinical software products, but a variety of approaches are applicable for solving communication issues between the systems.

The whole software development consists of central modules development, establishing the connection of the system to its environment, and specific modules development (see Figure 1). At the moment, HALMED has finished the first phase of system's development and the project is entering the second phase. The first phase dealt with the facilitation of HALMED's internal ADR processing and the development of Web reporting form and smart PDF form. The result of the first phase is a functional Web ADR reporting form and completed OPeN ingest functionality. The interface of the OPeN system is user-friendly and designed to both simplify and decrease the time required for the data entry. During the reporting, HCPs can select values in fields from various lists of medicinal products, institutions, units, and measurements. Users can also choose one of the existing templates for new reports.

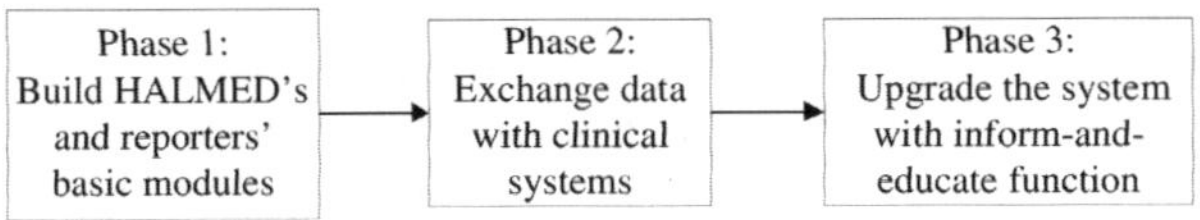

Figure 1. OPeN's system development phases.

B. *The OPeN system modules, functionalities and use*

HALMED has developed OPeN as a Web-based application that consists of a database, a Web interface, and Web services for integration with other IT systems (external clinical IT systems and other IT systems in HALMED). OPeN has been based on the Windows technology and developed in .NET Framework, while the data is stored in the Microsoft Structured Query Language server database (MS SQL). Firewall and Secure Sockets Layer (SSL) protocol protect the entire system. Access and data protection are established through the domain admin policies for internal users (Windows Active Directory) and the ones for external users. Backup procedures are performed regularly. HALMED's goal with this project was to automate reporting of adverse drug reactions by the HCPs directly via IT systems they use at their workplace. The scenario is the following: HCP submits reports and transfers them to HALMED, where the staff in charge of this task can process them (see Figure 2).

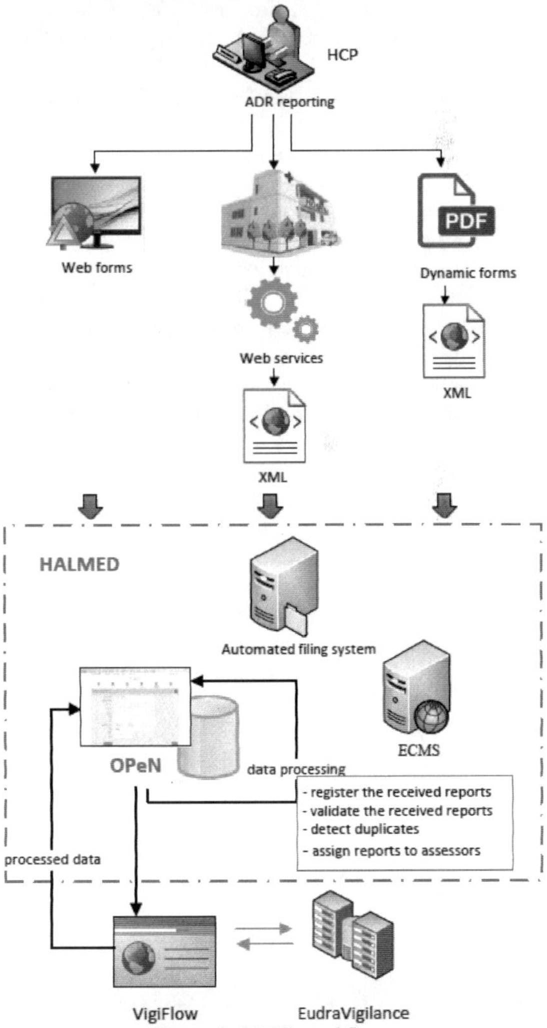

Figure 2. OPeN's workflow

The design of OPeN predicts connecting the system with CEZIH (Central Health Information System of the Republic of Croatia) and also with all IT systems used by HCPs, like hospital systems, general practitioners' IT systems, systems in pharmacies, private practices, and some additional stakeholders' systems. Most of the data required for reporting ADRs are already contained within HCPs' IT systems, so it takes less time to fill out the forms; it simplifies the process of sending data to HALMED as well. All things considered increase the quality of sent data, reduce the possibilities of errors during data entry, and prevent the submitting of the incorrect or incomplete forms.

The Web part of OPeN consists of two modules: the module for HALMED's employees and the module for external users (HCPs). Figure 3 shows OPEN's data entry interface (with creation dates and statuses - "Received" or "Task accepted"). HALMED's employees use the system to register the received reports, to validate them, to detect duplicates, to assign them to assessors, to track their processing, and any further steps (e.g., sending them to EudraVigilance system [23]). The module for HALMED's employees is also connected to HALMED's Enterprise Content Management System (ECMS) via Web services, where all the digital records are stored. These digital records are available for reading and editing in the OPeN system during the entire process. External users (HCPs) will use the system to report adverse drug reactions, having registered with the purpose of authentication.

All institutions that participate in data exchange will use the same list of HCP authentication data. This facilitates data exchange without the need for additional authentication at the moment of sending the data to HALMED. A common list will enable data traceability. Data exchanged between internal and external systems will be in XML data format. XML structures are following E2B-R2 and E2B-R3 formats [10] [11]. In the first phase of the project, HALMED has developed a smart PDF form which also includes embedded lists. The form is available via HALMED Website and OPeN system Website, currently in the test phase [24]. Anyone can access the form, fill it out and send it via e-mail. OPeN contains the function of automatic data loading from the form into the OPeN database, which eliminates the requirement to enter data manually.

Figure 3. OPeN's data entry interface

III. Discussion

A. *Efforts to increase the number and quality of ADRs in Croatia during the last decade*

Spontaneous ADR reporting is the keystone of pharmacovigilance. A sufficient number of high-quality ADR reports are directly correlated to the effectiveness of ADR systems' capacity in detecting drug safety issues. ADR systems are reliant upon the goodwill of HCPs and patients, not only on identifying suspected ADR reports, but also on reporting them. ADR reporting schemes are recognised to be subject to underreporting [25]–[27]. Directive 2010/84/EU, which came into force in 2013, identified the issue and stated that NCAs need to encourage and improve reporting of ADRs. According to the Article 102 of the Directive, NCAs shall "take all appropriate measures to encourage patients, doctors, pharmacists and other healthcare professionals to report suspected adverse reactions to the national competent authority" [28]. Heads of Medicines Agencies (HMA) within their *Strategy* for 2011-2015 have further propagated the aim of the Directive [29]. HMA aimed to support the strengthening of spontaneous ADR reporting systems, indicating four main complementary approaches as a mean to improve ADR reporting and achieve a more robust national pharmacovigilance system: (1) education, (2) motivation, (3) promotion, and (4) facilitation of ADR reporting, all of them mainly directed to HCPs and patients.

Throughout the last decade, HALMED has been included in numerous activities to increase the number and quality of reported ADRs in Croatia. The number of spontaneous ADR reports increased from 856 in 2007 to 3486 in 2016, bringing Croatia into top ten countries in the world per number of ADRs per million inhabitants [30]. Education, motivation, and promotion have been extensively used and combined with different ways of facilitation of reporting. The facilitation was done by introducing novel methods for capturing ADRs: online reporting form for patients and mobile app for patients and HCPs. On-line reporting form was launched in 2012: by 2016, 20% of all ADRs were received through this channel. Although HALMED intended the on-line reporting form for patients, it was widely used by HCPs. The PhV mobile app was introduced in 2016 (through project WEB-RADR) and it accounted for the 2% of ADR received in 2016. The focus was on activities aimed at facilitation of reporting. HALMED started introducing electronic reporting and it influenced the rationalisation of internal ADR processing. The main reason was a reduction of manual entry into the national ADR database. This issue gained more importance as the increasing number of received ADRs began to have a real impact on resources available to other pharmacovigilance processes within HALMED (e.g., assessments of risk management plans, periodic update safety reports).

B. Direct reporting from clinical systems as advancement in the field of pharmacovigilance

Although regulatory workload in HALMED increased substantially from 2013 onwards, there was still an active commitment to improving the national PhV system, as well as possibilities to detect signals and potential safety issues. This was this reason for planning and starting the OPeN project. OPeN aims to automate the internal handling of reported ADRs and to further increase both the number and the quality of ADR reports. The way to do it is by capturing ADR reports from online reporting form and clinical systems. The idea behind OPeN was to build a system that will allow the communication of clinical IT systems and the National ADR database. HALMED considers the capturing of ADR reports from clinical systems in electronic form to be the most challenging part of the project. Direct reporting from clinical systems, however, has many advantages. It improves reporting efficacy of HCPs by reducing their efforts to complete forms as it uses data from patients' records and automates data entry methods. HALMED considers it as a way of promotion of ADR reporting in general. According to the systematic review and meta-analysis published by I. Ribeiro-Vaz et al. [31], projects that aim to promote ADR reporting by using IT represent an increasing trend. According to their aggregate analysis, these interventions doubled the number of captured ADR reports.

Reporting from clinical system helps obtain complete information on ADRs. It significantly facilitate case assessment by providing context for ADRs. An additional value is the possibility to prompt reporters to complete a report within the system when specific tasks are completed, e.g., drug withdrawal from therapy. It can also prompt a reporter to enter data that are more particular into the report (e.g., a batch number in case of ADR related to vaccines and biologicals). This information is rarely captured in paper ADR form; it can, however, be crucial for understanding the cause of the ADR.

An additional benefit of automatic data entry method in direct ADR reporting from a clinical system is the use of controlled data entry. Where applicable, ADR reporters could use data registries (e.g., medicinal product registry) to facilitate their data entry method. Another benefit of using IT is the option to use data validation rules and to facilitate receiving accurate and valid reports. In this way, national ADR databases are becoming more accurate too. It reduces the need to contact the reporter in case of non-valid ADRs. Rules of reporting should meet the applicable ICH E2B standards [10] [11] and minimum required data. Additional validation rules should also be built in if needed to meet business or regulatory needs. Besides, the use of dictionaries, including the mapping of dictionaries, should help set up standardised terminology and coding terms for assessors. The use of dictionaries would represent a significant benefit for any pharmacovigilance system. The second phase of OPeN will address these issues.

C. A short comparison of pharmacovigilance systems with mechanisms of capturing ADRs from clinical systems

According to the SCOPE results, there are only 2 NCAs that use the possibility of capturing ADRs from clinical systems: regulatory agencies from Spain and UK (Agencia Española de Medicamentos y Productos Sanitarios – AEMPS, Medicines and Healthcare products Regulatory Agency - MHRA) [32] [33]. Their systems are mainly set to retrieve ADRs from General Practitioners (GPs), and this decision is logical since they have access to most of the patients' health information. Although their systems differ, their experiences indicate that the integration of ADR reporting with electronic health records, primary care, and e-prescription systems is positively correlated with the increased quantity of ADR reports and the quality of information received. In the UK, for example, during the pilot phase in 2011, automatic ADR reporting feature was added to one primary care system for GPs (SystemOne). Analysis of received Yellow Cards showed an increase for GP ADR reports of almost 50% compared to 2010. The UK and Spanish systems have a similar concept. There are some differences between the systems: ADRs received from clinical systems in Spain are not automatically uploaded into the national database Farmacovigilancia Española, Datos de Reacciones Adversas (FEDRA), but have to pass through their regional PhV centers. Also, the ADR description is in the form of free text, while the UK system allows ADRs to be coded using *The Systematized Nomenclature of Medicine* (SNOMED) dictionary [34]. The MHRA has built up a mapping between SNOMED concept terms and *Medical Dictionary for Regulatory Activities* (MedDRA) [35] which is standard terminology used for coding ADRs worldwide. It enables ADR reports to be automatically loaded into the MHRA's PV database without the need to code ADRs. OPeN will allow direct upload of ADR reports to the national database, however, coding of ADRs by the reporter will not be possible.

HCPs in Croatia use International Statistical Classification of Diseases and Related Health Problems 10 (ICD 10) terminology for coding medical terms [36] and no mapping to MedDRA is available. In addition to this, OPeN will allow for two-way communication with the reporters, mainly for retrieving follow-up information. We believe this is the benefit of our system, although no information on the communication possibilities within the UK or Spanish systems is widely available. There is limited published information on the particular data elements of both systems; we were able to compare our systems solely by actively participating in the SCOPE project. However, the information on the systems might be outdated since the SCOPE includes data up to 2013. Also, it is challenging to detect the publication on the efforts of other countries in implementing ADR reporting from clinical systems. As per our knowledge, Netherlands is piloting such a system. Previous research of the systems outlined above helped HALMED's team to avoid straggling in the development, implementation, and fine-tuning of the OPeN system.

IV. Conclusions and Expected Benefits

Submissions of timely and well-completed ADR reports increase the data available to NCAs and their capacity to protect public health in the end. It improves NCAs' ability to detect, identify, investigate, and act on potential drug safety issues.

Pharmacovigilance professionals consider the integration of clinical IT systems and ADR databases as bleeding edge tool for increasing both the number and the quality of captured ADRs. HALMED initiated OPeN for this very reason. Croatian NCA started with the initiative that aimed for the integration of the national ADR database, CEZIH, and various hospitals IT systems. Expected benefits include:

- Facilitation and promotion of ADR reporting,
- A higher number and better quality of reports,
- Reduction of manual data entry and data processing time per report,
- Standardisation, clarification of professional terms, dictionaries,
- Introduction to analytics, advancement in tracking Key Performance Indicators (KPI) and signals, planning for BI features,
- Sophisticated and proactive protection of public health.

V. Further Work

An efficient and connected health system and its components facilitate the protection of patients, efforts of HCPs, and beneficial activities of all other stakeholders in the national health domain. The OPeN system is an excellent example of the vital technological element of the national health system. Although Web-application OPeN is fully functional, HALMED is working on signing the protocol with the Croatian National Institute of Public Health to enable data exchange between these key health sector institutions. The system can become fully operational on a national level only after strengthening this cooperation. In the meantime, the Ministry of Health of the Republic of Croatia inserted activities related to OPeN into one of their strategic goals for 2019. Besides, concurrently with the signing of data protocol and with preparing for the second phase, HALMED is planning the third phase (see Figure 1) or system upgrade with the additional module that will cover informing and online education of HCPs.

References

[1] Agency for Medicinal Products and Medical Devices of Croatia. *About HALMED.* [Online], Available from: http://www.halmed.hr/en/O-HALMED-u/ [last access: 03/01/2018]

[2] European Medicines Agency. *Pharmacovigilance.* [Online], Available from: http://www.ema.europa.eu/ema/index.jsp?curl=pages/regulation/general/general_content_000258.jsp [last access: 01/30/2018]

[3] I. R. Edwards and J. K. Aronson, "Adverse drug reactions: definitions, diagnosis, and management," The Lancet, vol. 356, issue 9237, pp. 1255 – 1259, Oct. 2000.

[4] N. Mirosevic Skvrce et al., "Adverse drug reactions caused by drug-drug interactions reported to Croatian Agency for Medicinal Products and Medical Devices: a retrospective observational study," Clinical Sciences, vol. 52, issue. 5, pp. 604-614, Oct. 2011, doi: 10.3325/cmj.2011.52.604

[5] L. Hazell and S.A.W. Shakir, "Under-reporting of adverse drug reactions: a systematic review," Drug Saf. 2006; vol. 29, issue. 5, pp. 385-396, May 2006, doi:10.2165/00002018-200629050-00003

[6] European Medicines Agency. Information Management Division. *Master Data Management Roadmap: Substance, Product, Organisation and Referential data*, 23 April 2015. [Online]. Available from: http://www.ema.europa.eu/ema/pages/includes/document/open_document.jsp?webContentId=WC500186290 [last access: 01/30/2018]

[7] European Medicines Agency. *Monitoring of medical literature and entry of adverse reaction reports into EudraVigilance.* [Online]. Available from: http://www.ema.europa.eu/ema/index.jsp?curl=pages/regulation/general/general_content_000633.jsp&mid=WC0b01ac05808ce84c [last access: 01/30/2018]

[8] European Medicines Agency. *Reporting requirements for marketing-authorisation holder.* [Online]. Available from: http://www.ema.europa.eu/ema/index.jsp?curl=pages/regulation/general/general_content_000594.jsp& [last access: 01/30/2018]

[9] ICH, FDA, Center for Drug Evaluation and Research (CDER), Center for Biologics Evaluation and Research (CBER). *Guidance for Industry - E2B(R3) Electronic Transmission of Individual Case Safety Reports (ICSRs) Implementation Guide – Data Elements and Message Specification*, February 2014.

[10] ICH E2B(R2). *Individual Case Safety Report (ICSR) Specification and Related Files.* [Online]. Available from: http://www.ich.org/products/electronic-standards.html [last access: 01/30/2018]

[11] ICH E2B(R3). *Individual Case Safety Report (ICSR) Specification and Related Files.* Available from: http://www.ich.org/products/electronic-standards.html [last access: 01/30/2018]

[12] M. Liu, M.E. Matheny, Y. Hu, and H. Xu, "Data Mining Methodologies for Pharmacovigilance", ACM SIGKDD Explorations Newsletter, vol. 14, issue 1, pp. 35-42, June 2012, doi: 10.1145/2408736.2408742

[13] A. M. Wilson, L. Thabane, and A. Holbrook, "Application of data mining techniques in pharmacovigilance," British Journal of Clinical Pharmacology, vol. 57, issue 2, pp. 127–134, Feb. 2004, doi:10.1046/j.1365-2125.2003.01968.x

[14] European Medicines Agency. *Substance, Product, Organisation and Referential (SPOR) master data.* [Online]. Available from: http://www.ema.europa.eu/ema/index.jsp?curl=pages/regulati

on/general/general_content_000645.jsp&mid=WC0b01ac058078fbe2 [last access: 01/30/2018]

[15] ISO/HL7 27953-1: 2011 Health informatics – Individual Case Safety Reports (ICSRs) in pharmacovigilance – Part 1: The framework for adverse event reporting

[16] ISO/HL7 27953-2: 2011 Health informatics – Individual Case Safety Reports (ICSRs) in pharmacovigilance – Part 2: Human pharmaceutical reporting requirements for ICSR

[17] J. Almenoff et al., "Perspectives on the Use of Data Mining in Pharmacovigilance," Drug-Safety vol 28 (11), pp. 981-1007, 2005, doi:10.2165/00002018-200528110-00002

[18] T. Saracevic, "Relevance: A Review of and a Framework for Thinking on the Notion in Information Science," Journal of American Society for Information Science, vol.26, issue 6, pp. 321-343, Nov. 1975, doi: 10.1002/asi.4630260604

[19] SCOPE Work Package 4 ADR Collection, *IT Systems for ADR Reporting: Best Practice Guide* [Online], Available from: http://www.scopejointaction.eu/_assets/files/WP4-IT-Systems-for-ADR-Reporting-BPG.pdf [last access: 01/30/2018]

[20] IMI. *WEB-RADR: Recognising Adverse Drug Reactions.* [Online]. Available from: https://web-radr.eu/ [last access: 01/30/2018]

[21] SCOPE Work Package 4 ADR Collection, *Increasing Awareness of National Adverse Drug Reaction Reporting Systems: Best Practice Guide.* [Online], Available from: http://www.scopejointaction.eu/_assets/files/WP4.3-Survey-Report-layout-draft-4-final.pdf [last access: 01/30/2018]

[22] Central Health Information System of the Republic of Croatia (CEZIH). [Online], Available from: http://www.cezih.hr/ [last access: 03/01/2018]

[23] European Medicines Agency. *EudraVigilance.* [Online]. Available from: http://www.ema.europa.eu/ema/index.jsp?curl=pages/regulation/general/general_content_000679.jsp&mid=WC0b01ac05800250b5 [last access: 01/30/2018]

[24] HALMED. *Open (v.0.2.66.0)* [Online], Available from: http://open-test.halmed.hr/client/#/login [last access: 01/30/2018]

[25] Y. Moride, F. Haramburu, A. A. Requejo, and B. Begaud. (1997) "Under-reporting of adverse drug reactions in general practice." British Journal of Clinical Pharmacology, vol. 43, issue 2, pp. 177-181, Feb. 1997, doi: 10.1046/j.1365-2125.1997.05417.x

[26] R. M. Martin, K. V. Kapoor, L. V. Wilton, and R. D. Mann, "Underreporting of suspected adverse drug reactions to newly marketed ("black triangle") drugs in general practice: an observational study." BMJ, vol. 317, issue 7151, pp. 119-120, July 1998. [Online]. Available from: http://www.bmj.com/content/317/7151/119 [last access: 01/30/2018]

[27] Hazell L. and Shakir SAW. (2006) "Under-Reporting of Adverse Drug Reactions A Systematic Review." Drug Safety 29(5): 385-39

[28] Official Journal of the European Union. *Directive 2010/84/EU of the European Parliament and of the Council* [Online], Available from: https://ec.europa.eu/health/sites/health/files/files/eudralex/vol-1/dir_2010_84/dir_2010_84_en.pdf [last access: 01/30/2018]

[29] Heads of Medicines Agencies. *A Strategy for the Heads of Medicines Agencies, 2011-15.* [Online], Available from: http://www.hma.eu/fileadmin/dateien/HMA_joint/02-_HMA_Strategy_Annual_Reports/02-HMA_Strategy_Paper/2010_12_HMA_StrategyPaperII.pdf [last access: 01/30/2018]

[30] HALMED. *Adverse Drug Reaction Report for 2016* [Online], Available from: http://www.halmed.hr/fdsak3jnFsk1Kfa/ostale_stranice/Izvjesce-o-nuspojavama-za-2016.pdf [last access: 01/30/2018]

[31] I. Ribeiro-Vaz, A. M. Silva, C. Costa Santos, and R. Cruz-Correia, "How to promote adverse drug reaction reports using information systems – a systematic review and meta-analysis." BMC Medical Informatics and Decision Making, vol. 16, issue 27, Mar 2016, doi:10.1186/s12911-016-0265-8 [Online]. Available from: https://www.ncbi.nlm.nih.gov/pmc/articles/PMC4772685/# [last access: 01/30/2018]

[32] Agencia Española de Medicamentos y Productos Sanitarios (AEMPS). *Portada La AEMPS.* [Online]. Available from: https://www.aemps.gob.es/laAEMPS/portada/home.htm [last access: 03/01/2018]

[33] Medicines and Healthcare products Regulatory Agency (MHRA). *About us.* [Online]. Available from: https://www.gov.uk/government/organisations/medicines-and-healthcare-products-regulatory-agency/about [last access: 03/01/2018]

[34] SNOMED International. *SNOMED CT.* [Online]. Available from: https://www.snomed.org/snomed-ct [last access: 01/30/2018]

[35] MedDRA MSSO. *MedDRA. Medical Dictionary for Regulatory Activities.* [Online]. Available from: https://www.meddra.org/ [last access: 01/30/2018]

[36] eHDSI MVC. *WHO ICD-10 The International Statistical Classification of Diseases and Related Health Problems 10th Revision.* [Online]. https://ec.europa.eu/cefdigital/wiki/display/EHSEMANTIC/WHO+ICD-10+The+International+Statistical+Classification+of+Diseases+and+Related+Health+Problems+10th+Revision [last access: 01/30/2018]

Interactive Data Exploration Supporting Elderly Care Planning

Harry F. da Cruz*, Marie Gebhardt†, Felix Becher** and Matthieu-P. Schapranow*
*Hasso Plattner Institute, Digital Health Center,
Rudolf-Breitscheid-Str. 187, 14482 Potsdam, Germany
{harry.freitasdacruz | schapranow}@hpi.de
†Data Experts GmbH, Allee der Kosmonauten 33G, 12681 Berlin, Germany
marie.gebhardt@data-experts.de
**Gesundheitswissenschaftliches Institut Nordost der AOK Nordost, Wilhelmstr. 1, 10963 Berlin, Germany
felix.becher@nordost.aok.de

***Abstract*—Over the past decades, improvements in healthcare resulted in longer life expectancies and growing demand for specialized long-term care services for the elderly. Detailed up-to-date indicators describing the demographic status quo in order to enable adequate planning to meet the requirements of this population are missing. Today, these data are either scattered across different sources or only available in highly aggregated form. In the given paper, we share details about our interactive data exploration tool, which incorporates routine data from a large German health insurance to derive specific elderly care indicators, while ensuring scalability, data actuality and separation of data and visualization. Thus, governmental experts can access up-to-date data for elderly care planning in a structured and harmonized way.**

***Keywords*–Elderly care planning; key indicators; data exploration; on-line analytical processing.**

I. INTRODUCTION

Life expectancy has been increasing steadily in recent decades, particularly in high-income countries [1]. Coupled with declining birth rates, industrialized nations are faced with an ever-older population [2]. Competent authorities need to plan accordingly in order to meet changing societal needs, e.g., in terms of employment, pension funds, and health care provision. Although this is a widespread phenomenon in the developed world, Germany's position is particularly unfavorable: it ranks as the country with the oldest population in Europe and the second oldest worldwide [3].

While data on demographical developments in the form of census are usually available, data concerning elderly care, for instance the number of dementia patients in a given municipality, are scarce and/or insufficient, often lacking in quality and spread across different sources. This fact substantially hampers governmental planning efforts to this extent. Nevertheless, access to such data is crucial in devising adequate public policies that meet the needs of the elderly. In particular, a growing trend can be ascertained to support 'aging in place', that is, close to one's home and/or community, which is widely considered a more desirable alternative when compared to institutional care [4].

Healthcare data, e.g., in insurance companies can help bridge these data gaps, providing the basis for informed decision making. Figure 1 depicts the individual components of our tool modeled as a Fundamental Modeling Concepts (FMC) block diagram [5]. It builds on a combination of de-identified routine data from a large German health insurance together

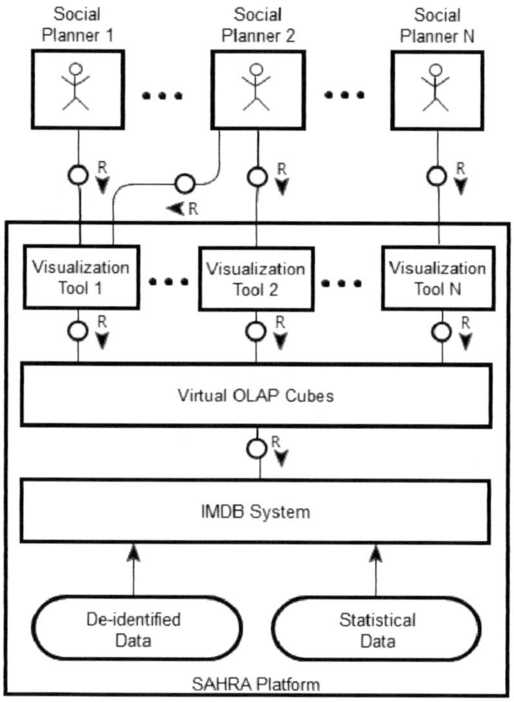

Figure 1. Software system architecture of the elderly care data exploration tool depicted as a FMC block diagram.

with statistical data provided by governmental bodies, both of which can be interactively explored by expert users. Data is integrated and harmonized in an In-Memory Database (IMDB), which provides On-line Analytical Processing (OLAP) cubes for a number of visualization tools. Governmental users, i.e., social planners, then can use such tools to explore the available data. Our tool is part of the project Smart Analysis Health Research Access (SAHRA) [6]. The consortium consists of industry and academia partners who aim to explore the opportunities inherent to data generated by healthcare actors to address research questions relevant for policymaking and health care provision.

The remainder of this work is structured as follows: Section II introduces our motivation, and objectives of our tool. In Section III, we set our work in the context of existing initiatives concerning elderly care planning, whilst our software architecture, incorporated data sources and data visualization tools are outlined in Section IV. Software requirements are detailed in Section V. Our contributions are shared in Section VI and their impact is discussed in Section VII. Our work concludes with an outlook in Section VIII.

II. Motivation

In the following, we present the motivation and the objectives pursued within the development of the elderly care data exploration tool.

In Germany, municipalities are often aggregated into districts for administrative purposes. Districts serve therefore as an intermediate governmental entity between the German states and the municipal governments themselves [7]. However, information on elderly and nursing care needs from the census and other sources is often provided not at the municipality level, where facilities and personnel are needed, but rather at an aggregated level (district or state) - if available at all. This information gap hinders planning efforts carried out by districts, which must rely on incomplete, partial information or commission expensive data collection from private entities. Furthermore, the data collection process and subsequent generation of reports is often resource-intensive and time-consuming, since it takes place manually and data sources are scattered in different providers.

Health and nursing care insurance companies possess a wealth of detailed routine information on their insured patients, e.g., where they live, what services they have used, relevant diagnoses, demographics, how long they have been cared for, if they are in an inpatient or outpatient facility, and so forth. The opportunity to bridge the existing data gap by the secondary use of routine data was the starting point for the elderly care data exploration tool within the SAHRA consortium.

The overall objective of our proposed solution is to support the planning and decision-making process concerning elderly and nursing care needs within German districts and municipalities. To this end, relevant elderly care indicators, such as number of inhabitants with nursing needs, percentage of individuals in an in-patient facility, among others, needs to be identified. For the pilot project, two administrative districts in Northeastern Germany have been selected. They provided the requirements for the solution and subject-matter experts were available for feedback. The names of the respective districts will not be disclosed for data protection reasons. Finally, it was necessary to ensure that data extraction and analysis are carried out in an automated and privacy-compliant fashion. As such, the specific objectives of the solution presented in this paper are 1) define a set of relevant care indicators, 2) extract and anonymize pertaining routine insurance data and 3) develop a web platform for interactive exploration of the defined indicators with automatic data flows and compliance with privacy regulations. In Section IV, we provide specific details on how these objectives have been achieved.

III. Related Work

Research on Information and Communication Technology (ICT) tools used for elderly care is often focused on supporting care delivery processes, for example via collaboration tools and data exchange. Since a patient will be treated by a number of different professionals across the care continuum, it is vital that they cooperate effectively [8]. Technologies for Ambient Assisted Living (AAL) and Building Automation (BA) represent another line of active research [9].

In contrast, our tool is targeted at providing high-quality data to support policy and decision making. For effective policies to be devised, a prerequisite is access to high-quality, up-to-date, precise information. Initiatives to this extent include information portals put together by governmental statistics services, for instance, the German Statistical Office's elderly care indicators [10] or the Healthy Aging Data Portal by the Center for Disease Control [11]. Likewise, the National Board of Health and Welfare in Sweden provides information on the state of its elderly population [12]. Moreover, a solution developed by a Portuguese software company provides operational management for nursing care homes on the cloud [13]. While it could potentially provide detailed aggregated information on elderly patients, it lacks the coverage necessary since nursing homes are only one of the actors in the care continuum [14].

Such initiatives have a number of weaknesses compared to our presented approach. Firstly, the content they provide is largely static and does not enable user-driven exploration. Exception to this rule is CDC's portal [11] and the PORDATA Contemporary Portugal Database on the elderly [15]. In both tools the user can interactively select indicators of interest and generate visualizations, albeit limited to few options.

Secondly, the scope of the indicators available on those platforms is rather limited, e.g., PORDATA offers four indicators while the CDC has nine. We have gathered more than 30 indicators of interest, which are extensible with ease. Third, the level of granularity of the current solutions is restricted either to the national or state-level, being therefore of reduced use for social planners in a district or municipality, for whom only fine-grained information is relevant. Forth, unlike existing solutions, the platform presented offers different stratification dimensions, e.g., according to legal criteria such as care level. Finally, automated data processing and compliance with privacy regulations ensure that up-to-date information is presented to the user once it was made available through the data provider.

IV. Methods

In the following, we share details about the incorporated research methodology. Section IV-A outlines the approach utilized for requirements engineering, while Section IV-B expounds upon the tool's technical infrastructure.

A. Requirements Engineering

We followed the design thinking methodology to create a user-centered prototype via interviews with subject-matter experts for elderly care planning [16]. In subsequent workshops, the team distilled the interview results into personas and user stories. Personas are a generic entity that represents a typical user of the solution. In particular, two personas have been identified. The first is the head of the district's social department, in charge of interfacing with numerous political entities and providing, among other tasks, guidance on elderly care issues. The second is a specialized officer or social planner, responsible to carry out data analyses and elaborate the elderly care plan, i.e., a comprehensive document containing an appraisal of the status quo on elderly care and a number of recommendations. User stories establish the aims pursed by the personas and often take the form of 'as persona

X, I want feature *Y* to achieve purpose *Z*'. This generated a catalog of user stories that has guided the implementation of the tool. They laid out the foundation for a set of functional and non-functional requirements as per the ISO/IEC/IEEE 24765:2017 [17], which are dealt with in detail in Section V.

B. IMDB Technology

One of the centerpieces of our tool is the incorporated In-Memory Database (IMDB) technology, which enables real-time data processing and analysis of the stored data [18]. We refer to IMDB technology as a toolbox of Information Technology (IT) artifacts to enable processing of data in real-time in the main memory of server systems [19]. Our incorporated IMDB backend has been proven to be an adequate tool providing real-time analysis features for big medical data [20]. In the following, we introduce selected building blocks of the IMDB technology incorporated in our SAHRA platform to enable real-time analysis of data relevant for interactive elderly care planning by governmental users.

1) Column-Oriented Data Layout: Most modern relational database systems fall into the category of transactional databases and store their data in a row-oriented format, i.e. all attributes of a record are stored in adjacent blocks [21]. This is advantageous when all data attributes of a single row have to be processed at once. In contrast, analytical database systems store and process data column-wise, i.e., all entries per column are stored in adjacent blocks, which is beneficial when only selected attributes of a data set are accessed [22]. The incorporated IMDB system supports both database layouts on a per table basis. Since user queries access only a subset of available attributes, in our tool tables are stored in columnar format. Thus, only a fraction of data needs to be processed.

2) Lightweight Compression: Lightweight compression refers to a data storage representation that consumes less disk space than its original pendant [19]. Storing data column-wise facilitates lightweight compression techniques, such as run-length encoding, dictionary encoding, and difference encoding [23]. Thus, the overall main memory footprint is reduced and the processing of data accelerated as existing CPU caches are used more efficiently. Since the elderly care exploration tool shall cover millions of inhabitants of a given region over many time periods, data size can quickly become an issue. For example, lightweight compression applied to our dataset results in an average compression of 5:1.

3) Partitioning: Our incorporated IMDB system provides vertical and horizontal partitioning [24]. The former addresses large database tables by splitting them up into multiple column-wise subsets that can be distributed across individual servers [25]. The latter handles large data sets by dividing them into smaller chunks of data row-wise, which supports parallel search operations and improves scalability [19]. Taking advantage of this approach, data belonging to different districts can be distributed and accessed across different server nodes on demand.

4) Multi-Core and Parallelization: Modern computer system architectures are designed to provide multiple CPUs with each of them having multiple individual CPU cores. For this capacity to be fully exploited, application execution must be parallelized, thereby resulting in maximum processing speed. Our IMDB system supports parallelization on various levels, e.g., parallelization both on inter and intra-operation level [19]. For instance, user queries spanning across multiple municipalities can be processed across different computing nodes, leading to faster response times.

C. On-line Analytical Processing Cubes

An OLAP cube is a multidimensional dataset containing different measures that can be aggregated/summarized across multiple dimensions [26]. In spite of their widespread adoption in Business Intelligence (BI) applications and data warehouse solutions, traditional OLAP cubes are often materialized, meaning they must be re-created every time new data is available [27]. In the IMDB cubes are virtual, allowing set operations, such as join, union, projections, etc., without the use of persistent, fixed aggregates [28].

V. SOFTWARE REQUIREMENTS

The software requirements were derived from interviews with subject-matter experts as outlined in Section IV-A. Following ISO 24765:2017, we were able to identify a set of functional and non-functional requirements, which will be related in detail in the following sections [17].

A. Functional Requirements

Functional requirements establish the set of functions that a system must deliver to the user in terms of inputs, processing and outputs, commonly expressed as "the system shall *do*" [29]. The functional requirements the elderly care tool must fulfill are:

- **F1: Data Privacy**. Since the tool is based on sensitive health data, appropriate measures shall be taken to ensure patients cannot be re-identified.
- **F2: Pre-defined Elderly Care Indicators:** A standardized list of elderly care indicators must be available in the tool covering the most relevant aspects for care planning.
- **F3: Fine-grained Exploration:** The proposed solution shall provide fine-grained, as well aggregated information using different aggregation dimensions, such as gender, care level and municipality.
- **F4: Access to Latest Data**. The tool shall employ automatic data flows from the provisioner to the analysis tool to ensure access to latest data at all times.
- **F5: User-defined Visualizations:** The tool shall enable the user to flexibly create his own reports using available elderly care indicators.
- **F6: Visualization Sharing:** The tool shall enable the user to share a given visualization with another user in digital form.
- **F7: Geographical Data Exploration:** The tool shall provide the user with the ability to generate cartographic visualizations based on the relevant administrative units.
- **F8: Long-term Forecasts:** The tool shall provide forecasts on all available elderly care indicators in 5 year time spans (e.g., for 5, 10, 15 and 20 years ahead). The forecasts must take into account demographic trends, as well as socioeconomic variables of interest.

B. Non-functional Requirements

Non-functional requirements are concerned with the overall properties of a system, as opposed to its functions, often written as "the system shall be" [30]. In the following, non-functional requirements of our tool are listed:

- **NF1: Ease of Use**. The user interface shall be made of modern, self-explainable interaction components, which are familiar to the user, ensuring ease of use.
- **NF2: Fast Response Time**. The tool shall be designed so that even querying large amounts of sets does not result in a response time higher than an empirical threshold of two seconds to ensure interactive user experience [31].
- **NF3: Device Independence**.The tool shall be accessible from different device classes, e.g. from desktop computers and mobile devices.
- **NF4: Ease of Extensibility of Indicators**. Given that user needs are constantly evolving, the set of elderly care indicators shall be easily extensible, without the need to shut-down the tool for updates.

VI. CONTRIBUTIONS

In the following, we share details about our software system architecture, incorporated data sources and set-up of data exploration tools, while pointing out how the requirements defined previously are met.

A. Software Architecture

The software system architecture of our tool is depicted in Figure 1. It is hosted within the SAHRA platform, which complies with pertinent data privacy regulations (F1) [6]. Governmental users responsible for elderly care planning in the districts, i.e., social planners have different data visualization tools at their disposal. A social planner can access one or more visualizations tools and created content can be shared with other users and user groups (F6).

1) Data Visualization Tools: Based on the infrastructure presented, different data visualization tools can be employed as long as they can consume IMDB OLAP cubes. We used a proprietary business intelligence suite for this purpose [32]. The suite makes extensive use of HTML5 and modern Javascript libraries, ensuring ease of use (NF1). Furthermore, it provides a web-based interface that is also accessible from mobile devices, being therefore device-independent (NF3).

2) Virtual OLAP Cubes: The use of virtual OLAP cubes ensures that as new data is available, analytical applications, such visualization tools, are updated in real time (F4). Another advantage provided by the use of virtual OLAP cubes is extensibility: new measures and dimensions can be added from new data sources or from calculations performed on existing measures (NF4). The analytical cubes were built upon relational database tables that were harmonized and integrated within the IMDB system.

3) IMDB System: The incorporated data sources, which cover de-identified routine data from the insurance company along with statistical data are integrated and harmonized within the IMDB. The building blocks discussed in Section IV-B namely column-based storage, lightweight compression, partitioning and parallelization ensure a fast response time for the tool even with a large data foundation (NF2).

TABLE I. Excerpt of elderly care indicators

Category	Indicator
Nursing care (NC)	Patients with NC needs Patients in a NC home Patients in home NC Patients receiving a NC allowance Patients using a nursing service
Respite care (RC)	Patients entitled to RC Patients who used RC Days with RC per patient
Day care (DC)	Patients entitled to DC Patients who used DC Days with DC per patient
Short-term care (SC)	Patients entitled to SC Patients who used SC Days with SC per patient
Dementia-related	Patients with dementia diagnosis Patients in nursing homes Patients in nursing services Patients cared for by relatives

B. Incorporated Data Sources

Based on routine data collected from insured patients from a given district, the health insurance partner extracted and consolidated a dataset of different data columns into a relational database table, each corresponding to a specific elderly care indicator, according to patient, reference year and a number of other attributes. This formed the standard list of elderly care indicators (F2). The resulting dataset is a sparse table indicating whether a given indicator applies for a specific patient or not. Furthermore, demographical statistics and prognostics compiled for the federal states to which the pilot districts belong were incorporated. They contain demographics by age segment (0-100) and gender for each of the municipalities in a district [33]. Routine data and statistical data can thereby be joined on year, municipality, age segment and gender.

C. Exploration Measures and Dimensions

Interactive exploration is enabled by a combination of measures and dimensions. Measures correspond to the elderly care indicators themselves. They cover, amongst others, number of people with nursing care needs, percentages of home, in-patient or out-patient care and aggregation by age group, gender and care complexity, offering fine-grained exploration options (F3). A partial list of the indicators identified is provided in Table I for illustration purposes.

The complete set of defined measures contains more than 30 indicators.The consolidated dataset also covers regional information, e.g., municipality of residence, that can be used to enable geographical exploration (F7). Dimensions encompass grouping attributes that can be arbitrarily combined to generate visualizations. Table II provides an overview of the exploration dimensions available on the platform.

D. Separation of Visualization from Data

To illustrate this characteristic of the tool, consider that a social planner in a district is interested to analyze age group distribution for dementia patients stratified according to different care levels (German: *Pflegestufen*). They are used

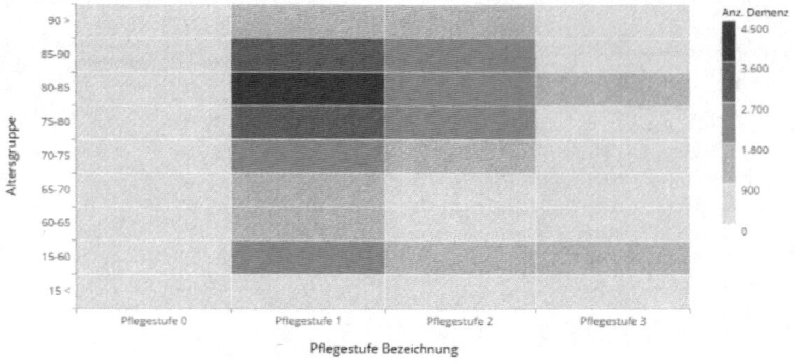

	Measures	Pflegestufe ...		
	Anz. Demenz			
Altersgruppe	Pflegestufe 0	Pflegestufe 1	Pflegestufe 2	Pflegestufe 3
65-70	78	1.059	684	137
70-75	234	1.952	1.233	493
75-80	437	3.286	2.294	867
80-85	331	4.195	2.682	963
85-90	198	3.011	2.254	847
90 >	34	1.201	1.388	540
Sum	1.669	17.741	12.387	4.949

Figure 2. Different alternative representation modes are available to the user. Image depicts age group distribution of dementia patients according to care levels: heat map (left) vs. tabular (right). With the platform, more time can be spent on interpreting data rather than manually building the required visualization.

TABLE II. Exploration dimensions

Category	Dimension
Patient-related	Age group Patient gender Dementia diagnosis Nursing dependent Patient dead
Regional	Federal state Municipal association Municipality District Municipality type Social region
Nursing care-related	Reference year Has nursing care needs Level of nursing care Grade of nursing care Type of nursing care

by insurance companies, among other purposes, to establish reimbursement guidelines [34]. By selecting the appropriate measure and dimensions on the platform via drag and drop, the heatmap depicted in Figure 2 (left) can be generated. It can be readily ascertained that dementia patients are more often assigned care level 1 (German: *Pflegestufe 1*), particularly after 70 years of age. If another mode of presentation is desired, the same information can be displayed in tabular form (right).

VII. Evaluation and Discussion

The tool developed meets the requirements put forth in Section V as follows. From a functional perspective, it provides a privacy-compliant framework (F1) and covers more than 30 pre-defined indicators (F2) and fine-grained aggregation criteria (F3). The data foundation is updated automatically, without a manual extraction, transform and load step, that is, as soon as new data are available (FR4). Furthermore, it allows users to create their own visualizations without any IT-knowledge (F5) and share them with other users (F6). In addition, the database was augmented with geographic coordinates to generate map-based visualizations (F7), not displayed here for the sake of brevity. Forecasting is provided by means of a linear regression of past indicators (F8).

Considering non-functional requirements, ease of use (NF1) is ensured by extensive use of HTML5 and modern Javascript libraries [35]. Besides, the use of the IMDB platform enables a fast response time even with a large underlying database (NF2), since data resides in the main memory [19]. Furthermore, the elderly care tool can be accessed from desktop and mobile devices alike (NF3). To support extensibility (NF4), the necessary calculations are performed using the IMDB's calculation views. They allow complex set operations, such as join, union, projections, etc., without the use of persistent, fixed aggregates, which must be re-created whenever requirements change [28].

Our tool is currently being evaluated in two administrative districts in Germany for user feedback. The tool has a number of advantages over existing portals offering statistics on elderly care, particularly in terms of its exploratory capabilities with different stratification dimensions, scope of the indicators available, level of information granularity (down to the municipality level), as well as automatic data flows within a privacy-compliant environment. As new districts are added to the tool, we expect a steep increase in data and user workload, which might impact system performance and response time, specially considering analytic queries. The building blocks of the IMDB technology discussed in Section IV-B such as lightweight compression, partitioning and multicore parallelization make it possible not only to use available resources more effectively, but also to add more computing power to the system without disrupting current operations, ensuring scalability.

However, from a technological standpoint, since the platform is based on proprietary technologies, it might pose a challenge for other institutions aiming to establish a similar solution. The principles and strategy which have guided the development of this tool can notwithstanding be utilized in a landscape based on open-source software tools, for example SpagoBI [36] or Pentaho with MySQL [37].

VIII. Conclusion and Future Work

We shared details of our software approach, the user requirements and the technical infrastructure that was developed. The data foundation is comprised routine healthcare data generated by the partner insurance company. The platform presents the user with a number of pre-defined measures and dimensions that allow for interactive exploration and generation of arbitrary visualizations.

Future work concerns three major aspects. The first aspect entails integrating available datasets with data generated across the entire care delivery continuum, including hospitals, nursing homes and out-patient services. Second, the tool currently provides linear regression forecasts based on the values of past indicators. However, this approach does not account for complex interactions between long-term socioeconomic and demographic variables. We aim to develop a more sophisticated forecasting model. Finally, user evaluation shall take place to assess the impact of the tool on current processes.

ACKNOWLEDGMENTS

Parts of the given work were generously supported by a grant of the German Federal Ministry of Economic Affairs and Energy (01MD15005).

REFERENCES

[1] C. D. Mathers, G. A. Stevens, T. Boerma, R. White, and M. Tobias, "Causes of International Increases in Older Age Life Expectancy," The Lancet, vol. 385, no. 9967, feb 2015, pp. 540–548.

[2] S. H. Preston and A. Stokes, "Sources of Population Aging in More and Less Developed Countries." Population and development review, vol. 38, no. 2, jun 2012, pp. 221–236.

[3] W. He, D. Goodkind, and P. Kowal, "An Aging World: 2015," in International Population Reports. US Census Bureau, 2016.

[4] N. L. Fields and H. Dabelko-Schoeny, "Aging in Place," in The Encyclopedia of Adulthood and Aging. Hoboken, NJ, USA: John Wiley & Sons, Inc., dec 2015, pp. 1–5.

[5] A. Knöpfel, B. Gröne, and P. Tabeling, Fundamental Modeling Concepts: Effective Communication of IT systems. J. Wiley & Sons, 2005.

[6] M.-P. Schapranow et al., "Towards an Integrated Health Research Process: A cloud-based approach," in 2016 Int'l Conf. on Big Data. IEEE, dec 2016, pp. 2813–2818.

[7] E. Pappermann, "Der Status der Gemeinden und Kreise als Gebietskörperschaften [The Status of Municipalities and Districts as Regional Entities]," in Handbuch der kommunalen Wissenschaft und Praxis. Berlin, Heidelberg: Springer, 1981, pp. 299–315.

[8] A. Duner, "Care Planning and Decision-making in Teams in Swedish Elderly Care: A Study of Interprofessional Collaboration and Professional Boundaries," Journal of Interprofessional Care, vol. 27, no. 3, may 2013, pp. 246–253.

[9] Y. Evchina and J. L. Martinez Lastra, "An ICT-Driven Hybrid Automation System for Elderly Care Support: A Rehabilitation Facility Study Case," Journal of Housing For the Elderly, vol. 30, no. 1, jan 2016, pp. 52–74.

[10] S. Bundesamt, "Pflege im Rahmen der Pflegeversicherung Deutschlandergebnisse [Care within Nursing Care Insurance, Germany Outcomes]," in Pflegestatistik 2015. Statistisches Bundesamt, 2017.

[11] CDC, "Healthy Aging Data," https://www.cdc.gov/aging/agingdata/index.html [retrieved: Feb 2, 2018], 2017.

[12] Socialstyrelsen, "Statistics on Care and Services for the Elderly 2015," http://www.socialstyrelsen.se/publikationer2016/2016-4-29 [retrieved: Feb 2, 2018], 2015.

[13] N. Loureiro et al., "A Web-based Platform for Quality Management of Elderly Care: Usability Evaluation of Ankira," Procedia Computer Science, vol. 64, jan 2015, pp. 666–673.

[14] R. Perley, Managing the Long-Term Care Facility: Practical Approaches to Providing Quality Care. Wiley, 2016.

[15] PORDATA, "Base de Dados Portugal Contemporâneo," http://www.pordata.pt/Portugal/Indicadores+de+envelhecimento-526 [retrieved: Feb 2, 2018], 2016.

[16] H. Plattner, C. Meinel, and L. Leifer, Design Thinking Research, ser. Understanding Innovation. Springer, 2012.

[17] International Organization for Standardization, "ISO/IEC/IEEE 24765:2017 - Systems and Software Engineering – Vocabulary," https://www.iso.org/standard/71952.html [retrieved: Feb 2, 2018], Geneva, Switzerland, 2017.

[18] P. Prassol, "SAP HANA als Anwendungsplattform für Real-Time Business," HMD Praxis der Wirtschaftsinformatik, vol. 52, no. 3, jun 2015, pp. 358–372.

[19] H. Plattner, A Course in In-Memory Data Management: The Inner Mechanics of In-Memory Databases. Springer, 2014.

[20] H. Plattner and M.-P. Schapranow, Eds., High-Performance In-Memory Genome Data Analysis: How In-Memory Database Technology Accelerates Personalized Medicine. Springer-Verlag, 2014.

[21] T. A. Halpin and T. Morgan, Information Modeling and Relational Databases. Morgan Kaufmann Publishers, 2008.

[22] D. J. Abadi, P. A. Boncz, and S. Harizopoulos, "Column-oriented Database Systems," Proceedings of the VLDB Endowment, vol. 2, no. 2, aug 2009, pp. 1664–1665.

[23] P. Svensson, "The Evolution of Vertical Database Architectures – A Historical Review," in Proceedings of the 20th Int'l Conf on Scientific and Statistical Database Management. Springer-Verlag, 2008, pp. 3–5.

[24] S. Lightstone, T. Teorey, and T. Nadeau, Physical Database Design: The Database Professional's Guide to Exploiting Indexes, Views, Storage, and more. Morgan Kaufmann, 2007.

[25] J. M. Hellerstein, M. Stonebraker, and J. Hamilton, Architecture of a Database System, Foundation and Trends in Databases. Now Publishers, 2007, vol. 1.

[26] S. Chaudhuri and U. Dayal, "An Overview of Data Warehousing and OLAP Technology," ACM SIGMOD Record, vol. 26, no. 1, mar 1997, pp. 65–74.

[27] C. Koncilia and R. Wrembel, Data Warehouses and OLAP: Concepts, Architectures, and Solutions. Irm Press, 2006.

[28] A. Pattanayak, "High Performance Analytics with SAP HANA Virtual Models," Journal of Computer and Communications, vol. 05, no. 07, 2017, pp. 1–10.

[29] C. Dickerson and D. N. Mavris, Architecture and Principles of Systems Engineering. Auerbach Publications, 2016.

[30] G. Koelsch, "Nonfunctional Requirements," in Requirements Writing for System Engineering. Berkeley, CA: Apress, 2016, pp. 107–149.

[31] J. Nielsen, Usability engineering. Academic Press, 1993.

[32] S. DuttaRoy, "SAP BusinessObjects BI Platform," in SAP Business Analytics. Berkeley, CA: Apress, 2016, pp. 55–69.

[33] Jens Rümenapp, "Kleinräumige Bevölkerungsprognose für den Landkreis Vorpommern-Greifswald [Small-scale Populational Forecast for the District Vorpommern-Greifswald]," Gertz Gutsche Rümenapp GbR, Tech. Rep., 2014.

[34] E. Gräßel, C. Donath, J. Lauterberg, C. Haag, and S. Neubauer, "Demenzkranke und Pflegestufen: Wirken sich Krankheitssymptome auf die Einstufung aus? [Dementia Patients and Care Levels: Do Disease Symptomes Affect the Classification?]," Das Gesundheitswesen, vol. 70, no. 3, mar 2008, pp. 129–136.

[35] J. Duckett, JavaScript and JQuery: Interactive Front-End Web Development, 1st ed. Wiley Publishing, 2014.

[36] M. Franceschini, "How to Maximize the Value of Big Data with the Open Source SpagoBI Suite through a Comprehensive Approach," Proceedings of the VLDB Endowment, vol. 6, no. 11, aug 2013, pp. 1170–1171.

[37] R. Bouman and J. van. Dongen, Pentaho Solutions: Business Intelligence and Data Warehousing with Pentaho and MySQL. Wiley, 2009.

An Ontology for Specifying Regulation-Compliant Genetic Privacy Policies

Michael Reep, Bo Yu, Duminda Wijesekera,
Department of Computer Science
George Mason University
Fairfax, VA, USA
e-mail: mreep@gmu.edu, byu3@gmu.edu, dwijesek@gmu.edu

Paulo Costa
Dept. of Systems Engineering and Operations Research
George Mason University
Fairfax, VA, USA
e-mail: pcosta@gmu.edu

***Abstract*— Genetic information provides important diagnostic data from patients to their health care providers and researchers that match phenotype and genotype. However, both diagnostic and research data providers must be confident that using this data for either purpose protects the data provider from foreseeable privacy breaches. In order to do so, Federal and State laws are in place to specifically address genetic information in addition to the laws established to protect generic health information. State genetic privacy laws diverge widely in their level of detail and constraints on releasing data, criteria for evaluating access to such data, data owner consents required to release data, and conditions for using released data. A rule-base specifying these variations can be used as a policy language to enforce data releases from electronic health records and gene pools. In order to satisfy this need, we describe a comprehensive ontology for genetic privacy based on existing applicable laws. Our ontology is used in ontological rule bases within medical workflows that are directly integrated with electronic health records. As shown in our ongoing work, this integration provides a solid foundation for enforcing laws and regulations in preventing unlawful disclosures of genetic information.**

Keywords- Genetic Privacy; Electronic Medical Records; Ontology; Health Care; Genomic Medicine.

I. INTRODUCTION

Patients are less likely to share data if there is a concern about privacy, so consents are necessary to help allay these concerns [1]. Privacy concerns have been heightened as Electronic Health Records (EHRs) have become widespread and ensuring privacy has increased in importance [2][3]. The privacy concerns that patients have about electronic medical records also apply to genetic information. There are demonstrable benefits to using genetic information as genetic studies map genotypic and phenotypic data directly to diseases, allowing for preventive and early interventional care to reduce morbidity and treatment costs [4][5]. These benefits have to be balanced against inherent unusual characteristics of genetic information that can identify a patient and his/her genetic relatives, therefore placing any of them at risk of negative consequences, such as discrimination [6]. Consequently, laws impose penalties if genetic data is inappropriately released. Studies have also shown that de-identification of genetic material may be insufficient to protect patient privacy [7][8].

In the United States, overall health privacy was addressed by the Health Insurance Portability and Accountability Act (HIPAA) of 1996, which was implemented to improve the efficiency and effectiveness of the US healthcare system. HIPAA was followed by the Privacy Rule in 2000 to address three covered entities: health plans, health care clearinghouses, and certain health care providers [9]. HIPAA was also followed by the Genetic Information Nondiscrimination Act of 2008 (GINA) to protect individuals from discrimination in employment and insurance based on genetic information [10]. Furthermore, almost every state and the District of Columbia have laws that specifically address genetic protections to some degree. Health Information Exchanges and direct sharing between health care providers are still subject to the applicable State laws even for interstate data transfers [11]. This paper develops an ontology that provides the syntactical elements (i.e., entities and their relationships) sufficient to specify applicable legislation and regulations in the forms of a structured formal rule-base.

In our previous work, we developed a prototype that uses a medical workflow system for an EHR to enforce Federal and State laws in addition to organizational policies. Workflows provided the mechanism to gather the necessary information within the context of an EHR. We prepared an initial genetic privacy ontology and sample rules to enforce laws in selected states to validate our approach.

Our next step is developing this comprehensive genetic privacy ontology based directly on relevant Federal and State laws. Following this Introduction, Section 2 addresses related work; Section 3 provides the methodology we followed in developing this ontology; Section 4 describes the genetic privacy ontology with detailed descriptions for each super-class; Section 5 provides an example of the ontology being used in our prototype; and, finally, Section 6 presents conclusions.

II. RELATED WORKS

There are existing standards and frameworks with methods to implement various aspects of genetic privacy protections. The Integrating the Healthcare Enterprise (IHE) standards profiling organization has developed frameworks, use cases, and specifications for managing the sharing of documents between organizations [12]. The inter-organizational policies must be completed prior to the use of this standard for implementing the consent agreements. There is some capability to address components of genetic privacy related to acknowledging consents but not all the required capability. For example, the use case of individuals specifying that other specific individuals do, or do not, have access to their data is listed as a scenario that is explicitly not supported [13]. Many State laws call for this type of consent specifications as a prerequisite for permissible access to data.

The restrictions placed by regulatory environments on information sharing has been identified as an issue that requires coordination across system silos [14]. The Global Alliance for Genomics and Health (GA4GH) provides a framework for sharing genome data with privacy and security policies, technology recommendations, guidance and architecture to allow interactions between organizations [15][16]. The basis of data sharing in GA4GH is that the donors or their representatives have provided consent in accordance with organizational policies and the applicable laws [17]. The work to date provides comprehensive policies but does not have a functional mechanism for implementing sharing data or addressing the restrictions placed by donors in systems that hold and use such data.

Other health-care privacy ontologies have some overlap with genetic privacy concepts based on laws. However, these ontologies have gaps in numerous areas when compared to State law implementation requirements. The HL7 Security and Privacy Ontology has a class *PurposeOfUseOntology* with a purpose code and description [18, p. 1]. Because the focus is on health care organizations, the main categories in this ontology are for health care marketing, operations, payment, research, public health and treatment with options for patient requested inquiries including family, power of attorney and support network. This list does not include key purposes regulated by law, such as Law Enforcement, Homeland Security and Insurance access. Other matching HL7 ontologies have some overlap (such as Organization, ObligationPolicy, Refrain, and Role) but not a complete set of genetic information related categories. The Sensitivity class contains a *genetic disease information sensitivity* but this needs to be set based on the state law attributes of the ontology. Many of the state laws have conditions that must be met prior to releasing genetic information in addition to imposing specific obligations to be adhered to after the release. A future research option is to develop a mapping and extension between our genetic focused ontology and the HL7 framework as a basis for an implementation.

Genetic privacy protections issues are expanding with the introduction of big data repositories and Direct-to-Consumer (DTC) DNA testing. Adoption of the latter has skyrocketed with its lower prices and wide-spread advertising. DTC DNA testing-related sites encourage sharing of genetic data, including through the use of social media. But consumers often do not have an understanding of the consequences of these services [19]. In general, even when presented with consent agreements, consumers, patients and research participants have a wide variety of reasons for permitting access to their data, do not always fully understand the extent and implications of these agreements, and underestimate the ability for de-identification [20][21].

Work by Rahmouni *et al.* developed an ontology of European privacy requirements for sharing patient data between countries [22][23]. It focused on the implementation of data access between countries with respect to privacy status, consent requirements, recipients, level of detail, purpose, secondary purpose, and access by legal representatives. There are no structures for the supplemental requirements prevalent in US laws outside various options for consent agreements and anonymization.

Other healthcare security focused ontologies lack the focus on purpose-driven access found in US laws. Blobel's pHealth has a policy structure that can implement many of the legal requirements and implements patient consent using policies [24]. The patient and internal organizational focus on access policies limits the opportunities to address the wide variety of scenarios prevalent with external access to patient data.

Most privacy models also use Role-Based Access Control (RBAC) to data inquiries and implementing enforcement policies. The use of RBAC has been identified as one of the candidates for implementing privacy access controls in the EHR domain [25], where rights can be assigned based on organizational policies in a hierarchical manner that is modified based on the user's role and then adjusted by the patient as desired. Healthcare privacy extensions, such as those proposed by Hung, provide the structure for adding concepts for areas including purpose, obligations, and retention [26]. The nature of genetic access restrictions and criteria requires a specific framework to accommodate the variations in State laws.

III. METHODOLOGY

A. Process

The goal of developing our genetic privacy ontology is to identify diverse factors relevant to enforcing these laws in the United States. The applicable Federal law specific to genetic privacy is the GINA [27]. For State laws, the National Human Genome Research Institute maintains a Genome Statute and Legislation Database [28]. This database classifies the laws into the following categories:

- Employment Nondiscrimination
- Health Insurance Coverage
- Health Insurance Nondiscrimination
- Other Lines of Insurance Nondiscrimination
- Other Topics
- Privacy
- Research
- Use of Residual Newborn Screening Specimens

A search against all categories generated a list of over 400 individual state statues that were reviewed for their applicability. The focus was for statues with criteria that would impact a request to an EHR or similar repository. References to other statues within those on the search list were followed when there was a potential for additional relevant use cases.

The next step in the process involved reviewing every relevant law for the specific terms and phrases associated with privacy protection. Statements and phrases related to the following super-classes (with two examples listed below for each item) were extracted:

- **Purpose**: What use case is being addressed? (law enforcement, treatment)
- **Subject**: Whose information is being protected? (individual, minor child, family member)
- **Requester Role**: what role is specified for individual making the request? (physician, genetic counselor)
- **Requester Organization**: What kind of organization does the requester represent? (insurance company, court)
- **Target of Request**: What kind of information or activity is protected? (genetic information, test results)
- **Pre-conditions**: What must be done before the information can be released? (obtain consent, approved by an institutional review board)
- **Post-conditions**: What must be done, or not done, once the information is provided? (non-discrimination, destroy after use)
- **Penalties**: How will any violations potentially be punished? (misdemeanor, fine)

Then, all super-classes were structured into relevant classes. For example, the Subject super-class was divided into the Individual, Immediate Family, Beneficiary, and Relative classes. The Relative class has additional subclasses for Blood Relative, Family Member, and Identical Brother. In order to ensure the appropriate coverage, similar terms were recorded in the detailed model (which is not presented here due to space constraints.) An example is that "Treatment" is used as the class name and is in the model, while "patient care" provides the same base meaning. Properties are added to the ontology solely based on the associated State law.

B. *Exclusions*

The expansive state database generated a comprehensive set of criteria for genetic privacy protection. Use cases not relevant to protecting genetic information in medical records were excluded from data elements analysis. For example, while there are State laws related to maintaining State DNA Databases for the criminal justice system, these laws were not evaluated after the first screening review. In addition, the search was not expanded to address Federal and State regulations as the basis of this guidance is derived from the associated laws. Finally, laws that provide generic protection to any component of protected health information and would be enforced across all information requests were not included. For example, while most State laws on genetic privacy are relatively recent and still being legislated, HIPAA restrictions have been in place since 1996 and are well established within the medical and insurance community [29].

IV. GENETIC PRIVACY ONTOLOGY

The Genetic Privacy Ontology as shown in Figure 1 is organized into four high-level components to reflect aspects of accessing medical records.

- **Requester** is information on the individual submitting the access request with their associated role and organization. The role and organization are linked with the purpose for a specific request. For example, a law may permit physician access to records for patient treatment at a hospital with a different set of conditions if the physician is participating in research at that facility.
- **Request** indicates characteristics of the person whose medical record is being accessed (subject), how the information will be used (purpose), what will be done with the information once received (action), and what information or activity is being requested (target).
- **Validation** provides conditions that must be addressed prior to information release (pre-conditions) and if a consent agreement is required with specific clauses to be included (consent).
- **Constraint** addresses limits placed on the use of the data (restrictions) and activities to be performed after the request is fulfilled (obligations).

This paper addresses the Requester and Request components as the most important aspects of enforcing the law. The laws vary widely in terms of which classes are specifically included and the amount of detail provided. As an example of a law with broad scope, the genetic information access law in Arkansas states *"Except as provided in (b) of this section ... (1) a person may not collect a DNA sample from a person ... unless the person has first obtained the informed and written consent of the person..."* followed by five specific purposes in section (b) where access is permitted.

A. *Purpose*

The Purpose super-class shown in Figure 2 provides the linchpin of genetic privacy protections. Every access to medical records must have a purpose (or reason for the access request). It is a violation of core security principles related to confidentiality to allow data access without a valid reason.

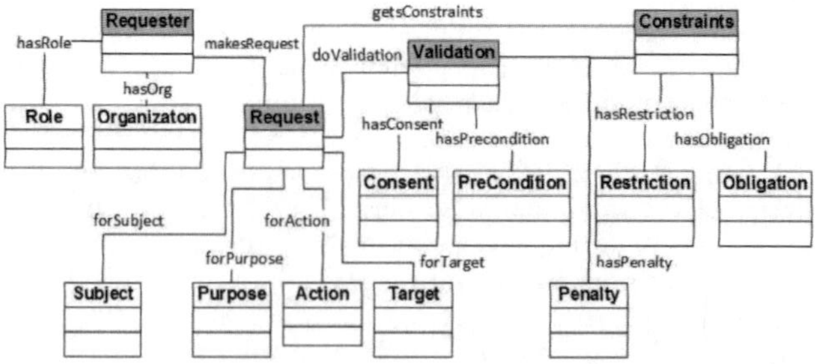

Figure 1. Genetic Privacy Ontology.

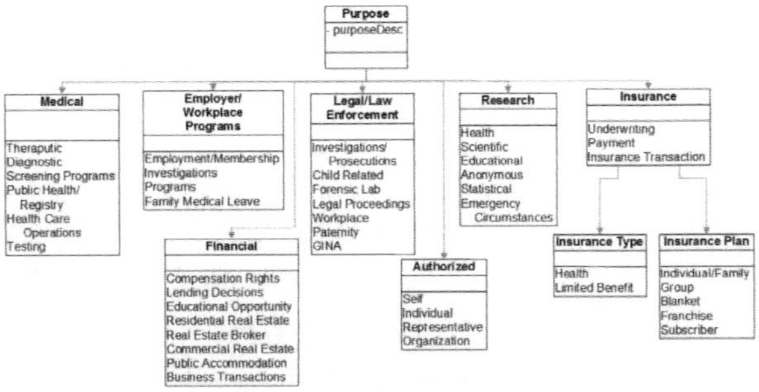

Figure 2. Purpose Super-Class.

The classes are:

- **Medical** with a focus on genetic-related activities plus access to genetic information for related efforts
- **Employer/Workplace Programs** includes labor organizations, apprenticeships, and licensing
- **Legal/Law Enforcement** for criminal, civil, court activities, compliance and related legal proceedings
- **Research** for health specific research, general scientific studies, educational programs, and access to deceased genetic information in an emergency situation
- **Insurance** for health plan underwriting, determining payments and other business transactions. There are two related subclasses for underwriting to determine the type of insurance being processed and the plan.
- **Financial** contains various financial transactions including whether a person is entitled to compensation for the use of their genetic material.
- **Authorized Person** allows access to records by people and organizations authorized in a consent agreement along with the individual themselves.

Many other classes have a direct relationship with a Purpose. For example, the Law Enforcement Purpose would only be applicable to Law Enforcement roles and other roles would be invalid. This linkage also applies to aspects of Organization and Target.

B. *Action*

During the full review of State laws, the need for an Action class as shown in Figure 3 became obvious to

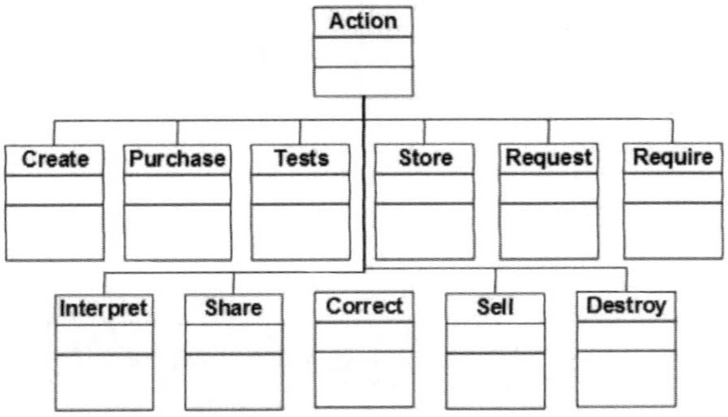

Figure 3. Action Super-Class.

indicate how the interactions with the medical record will occur. The list follows the information lifecycle of a medical record from acquisition through use and destruction. These terms are extracted directly from the laws and each class contain lists of similar terms. For example, *Require* also addresses the terms Inspect, Compel, and Order.

Request and *Require* are separated to reflect the ability to deny a request for information as opposed to an inquiry that indicates a demand based on a compelling reason. For example, there are restrictions as to when insurance companies may request information and that the information may not be required as a condition of underwriting. In some cases, both terms are used as seen in the District of Columbia's law stating that "*A health benefit plan or health insurer shall not request or require an individual or the individual's family member to undergo a genetic test.*"

C. *Subject*

The Subject super-class in Figure 4 is larger than the immediate person in an access request. This broad definition is needed in order to accommodate laws with statements like the definition from a Delaware law, "*'Genetic information' means information about inherited genes or chromosomes, and of alterations thereof, whether obtained from an individual or family member...*". The Individual has subclasses with specific terms that map to purpose. For example, the Legal subclass has a law enforcement attribute in the full model for "*individual has been convicted of a felony*" to address a State law in New Mexico.

The age data field is not only a numeric value but a conditional codifying the decision-making capacity and age ranges. Some of the options beyond years of age include "*child born in the state*" and un-emancipated minor. A child can be minor child or also a reference to an adult child in the Family Member subclass. Fetus and embryos are called out specifically in some laws so they are articulated as a Subject subclass. Dependents are presented as a separate subclass, as these individuals are not necessarily a family member. The same criteria holds for Beneficiary as the person may not be directly related to the individual. This subclass is included as

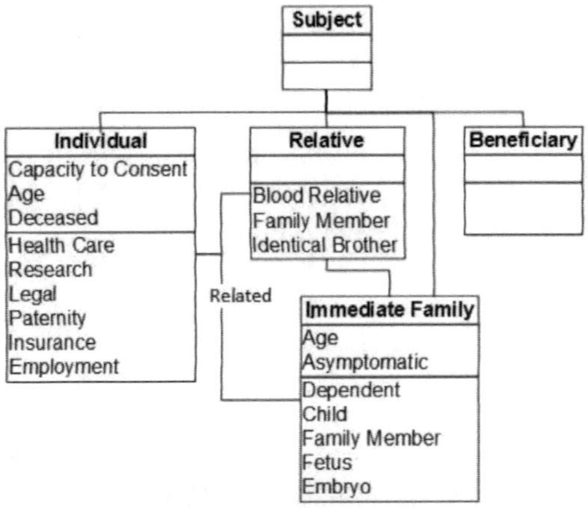

Figure 4. Subject Super-Class.

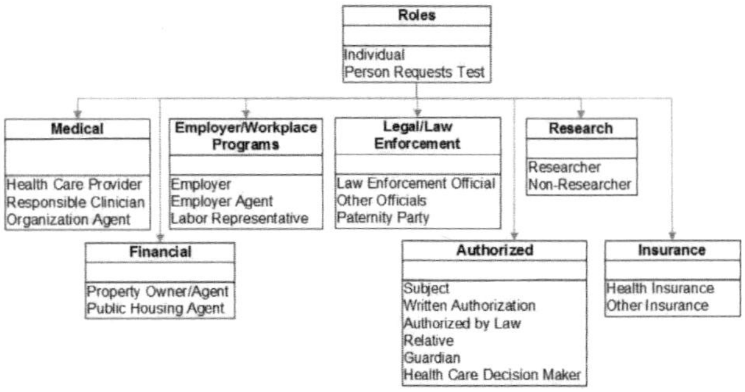

Figure 5. Role Super-Class.

Kentucky has a law that references the individual or their beneficiaries accessing genetic services.

D. Role

The majority of Roles in Figure 5 map to specific purposes and their use would be restricted to the associated purpose. There are general references to "*Individual*" accessing records so a broad subclass is needed for undefined people. Kentucky state law includes a reference to the person who orders a test on an infant or by the person registering the birth. The law is not clear that the request would be done by a health care provider so the category is separated.

Within an organization, the role becomes important for deciding on access. For example, a receptionist at a hospital does not have access to genetic information even though they work at a health care provider. Some terms are relevant to both individuals and organizations. For example, the term health care provider often means either entity.

E. Organization

Some terms that are typically associated with an individual may also be applicable to an Organization super-class which is provided in Figure 6. For example, the term *Person* can be a corporation in some states, such as defined in New Hampshire: "*Person includes a human being, an association or organization, a trust, corporation, and partnership*." The phrase "*Health Care Provider*" is also an organization in addition to a specific person/role. The use of subclasses with the same name as the Purpose super-class indicates a match where there are restrictions associated with the organization (and hence role). Therefore, a Research Organization would be limited to using Research Roles and Purposes for a valid combination.

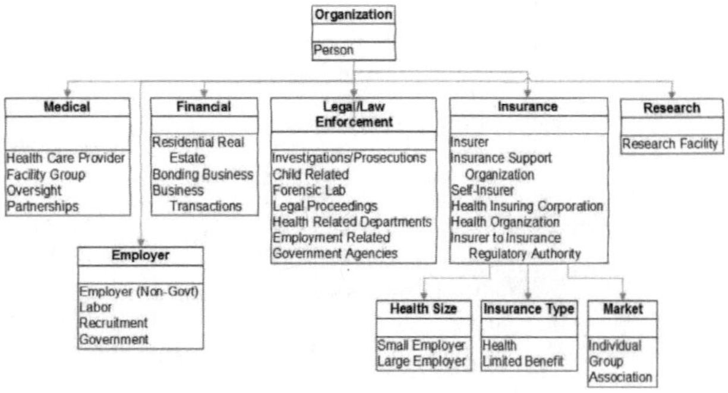

Figure 6. Organization Super-Class.

F. Target

The Target super-class in Figure 7 addresses the specific item being addressed in the law and encompasses more than genetic information.

- **Physical Specimens** are included as genetic material is derived from these sources and thus have specific access restrictions.
- **Record** includes all the information, data and audit records within an individual's medical records. Numerous states have restrictions not only on the results of genetic tests and activities, but also on whether a request was made and/or denied by an individual or family member. If a record has been de-identified, different handling is indicated in some states.
- **Organization** is included as some laws incorporate who is the custodian or holder of the records being accessed.
- **Health Status Information** (also called "health status related factors" and other phrases) is generally defined as including Health status, Medical condition (including both physical and mental illness), Claims experience, Receipt of health care, Medical history, Genetic information, Evidence of insurability including conditions arising out of domestic violence, and Disability. However, in some states genetic information is explicitly excluded as health status information.
- **Genetic Information** includes all the related subclasses. The state of Washington recently included genetic information as a biometric identifier so a state list is included.
- **Family Member** and their Genetic Information is often included in the scope of the individual's genetic information.
- **Research** addresses genetic information obtained in this purpose along with any genetic services obtained in association with the research. (This linkage is specifically called out.)
- **Genetic Services** is for those areas targeting genetic-

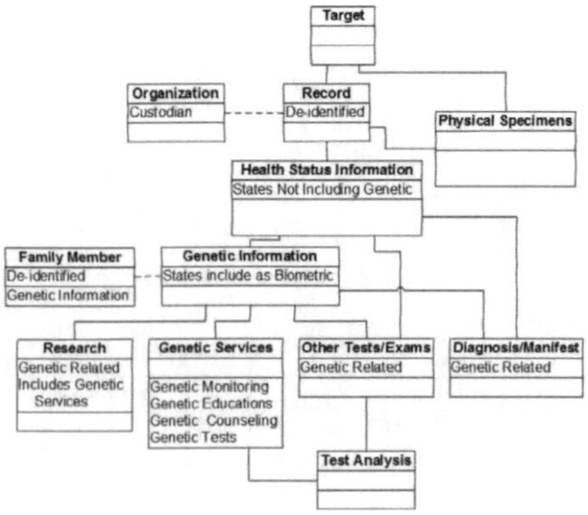

Figure 7. Target Super-Class.

related activities plus those tests performed specifically to identify genetic characteristics. The test results associated with these tests are also covered under the umbrella of protected genetic information.

- **Other Test/Exams** addresses medical procedures performed for reasons other than genetic services but the resulting information is genetic-related.
- **Diagnosis/Manifest** addresses those situations that indicate underlying genetic conditions not found directly by genetic services.

V. IMPLEMENTATION

Our previous prototype implemented a three-layer architecture for enforcing genetic information release criteria as seen in Figure 8. At the top layer, a workflow developed in Yet Another Workflow Language (YAWL) [30] orchestrates the information gathering on the Requester and Request, invokes the Consent Service layer, displays the access request decision (permit or deny) and gathers electronic signatures to enforce the Validations and Constraints. The Consent Service Layer uses Java code to obtain the information from the workflow, populate the ontology instances in Protégé, invoke the DL Reasoner, use the Rules Hierarchy Algorithm to combine the Federal, State and Local decisions into an overall final result, and populate the workflow variables with the results for display and action by the end user. The ontology itself in implemented in Protégé and the laws are encoded using Semantic Web Rule Language (SWRL) [31].

Our previous papers [32][33] provide extensive information on the prototype operation and detailed use cases. In this Section we provide targeted examples based on the purpose-focused ontology.

A. Related Organization

In New Mexico, genetic information access is allowed without patient consent *"(1) to identify an individual in the course of a criminal investigation by a law enforcement agency"*. The corresponding SWRL rule for this law is:

Rule: makesRequest(?r, ?req), inState(?req, "NM"), forResource(?req, ?resource), isGeneticResult(?resource, true), includesIdentity(?resource, true), forPurpose(?req, ?pur), isInvestigation(?pur, true), hasOrganizaton(?r, ?org), isLawEnforcement (?org, true), hasResponse(?req, ?resst), responseLevel(?resst, "State")→ isAllowed(?resst, true), canOverride(?resst, false), decisionSource(?resst, "NM LAW 24-21-3.C"),, hasRule(?resst, 3105)

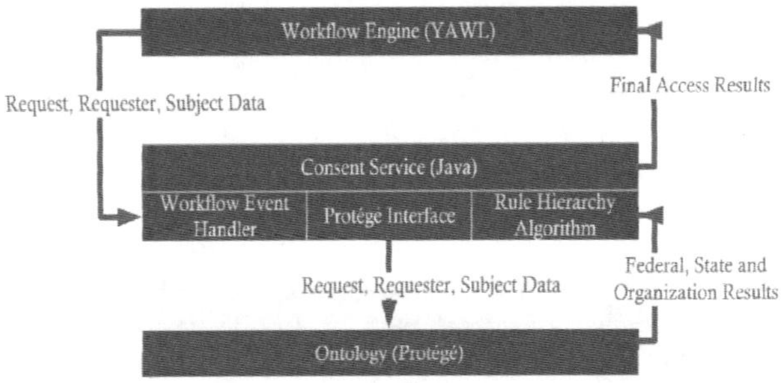

Figure 8. Prototype Architecture.

In this rule,

- ?r is for the Requester of the Request
- ?req is for the Request that links the various components, such as Subject, Purpose and Resource
- ?pur is the Purpose for the Request
- ?resource is for the "GeneticTestResults" part of the medical record
- ?org is the Organization f the Requester
- ?resst is the State Response object that is associated with the Request.

These SWRL statements are explained in Table I.

TABLE 1. SAMPLE PURPOSE-FOCUSED RULE

SWRL Statement	***Explanation***
makesRequest(?r, ?req)	Links Requester for the Request
inState(?req, "NM")	Request is for New Mexico
forResource(?req, ?resource)	Links Request with the Resource
isGeneticResult(?resource, true	Restricts the rule to a Resource that is identified as a genetic test results
includesIdentity(?resource , true)	Restricts the rule to Resources that are used to confirm identity
forPurpose(?req, ?pur)	Links Request with Purpose
hasOrganizaton(?r, ?org)	Links Organization with the Requester
isLawEnforcement (?org, true)	Confirms the Organization is a Law Enforcement Agency
hasResponse(?req, ?resst)	Links the Request with a Response to store answer
responseLevel(?resst, "State")	Gets the Response for the State level answers
-> isAllowed(?resst, true)	Sets the State response to access is allowed
canOverride(?resst, false)	Sets the State Response to not allow organization override
decisionSource(?resst," NM LAW 24-21-3.C ")	Sets the State response to reflect the decision source as state law
hasRule(?resst, 3105)	Sets the rule number to 3105 for reference

Boolean attributes are used in the ontology to simplify the evaluation of specific conditions. For example, *isLawEnforcement* allows any organization that meets the criteria to be provided access in relation to the *isInvestigation* attribute for Law Enforcement purposes. In numerous laws related to law enforcement, the statement specifically calls out using the genetic information for identifying a person. Therefore, an *includesIdentity* attribute is associated with the genetic information resources to exclude any that are medically focused. The Boolean attributes enhance the flexibility of the ontology implementation.

B. *Broad Statements*

Some states have broad statements that all uses for genetic information are denied (or permitted) except as outlined in a specific list. This scenario is addressed by assigning state-specific attributes to the *Purpose* class. Then the allowed *Purpose* instances are set to true and all other instances are set to false.

For example, in Alaska, consent is required to collect DNA sample, perform DNA analysis, retain DNA sample or results, or disclose results except for the following cases:

- Public Safety DNA database
- Law Enforcement purpose
- Determining Paternity
- Screen Newborns
- Emergency Medical Treatment

An attribute is added to *Purpose* for *isAKConsentRequired* which can be set to true for these specific instances. (The official state abbreviation for Alaska is AK.) Within the *Medical* class, most instances under the *Therapeutic* subclass would have attribute set to true with only Emergency Medical Treatment set to false. A sample SWRL rule that would be invoked for Emergency Medical Treatment is as follows:

makesRequest(?r, ?req), inState(?req, "AK"), forResource(?req, ?resource), isGeneticResult(?resource, true), forPurpose(?req, ?pur), isAKConsentRequired(?pur, false), hasResponse(?req, ?resst), responseLevel(?resst, "State") -> isAllowed(?resst, true), canOverride(?resst, false), decisionSource(?resst, "AK LAW 18.13.010"), hasRule(?resst, 23)

In this rule,

- **?r** is for the Requester of the Request
- **?req** is for the Request that links the various components, such as Subject, Purpose and Resource
- **?pur** is the Purpose that is associated with the Request
- **?resource** is for the "GeneticTestResults" part of the medical record
- **?resst** is the State Response object that is associated with the Request.

These SWRL statements are explained in Table II.

TABLE II. SAMPLE BROAD STATEMENT RULE – NO CONSENT

SWRL Statement	***Explanation***
makesRequest(?r, ?req)	Links Requester for the Request
inState(?req, "AK")	Request is for Alaska
forResource(?req, ?resource)	Links Request with the Resource
isGeneticResult(?resource, true	Restricts the rule to a Resource that is identified as a genetic test results
isAKConsentRequired(?pur, false)	Restricts the rule to Purposes that do not require consent
forPurpose(?req, ?pur)	Links Request with Purpose
hasResponse(?req, ?resst)	Links the Request with a Response to store answer
responseLevel(?resst, "State")	Gets the Response for State level to store answers
-> isAllowed(?resst, true)	Sets the State response to access is allowed
canOverride(?resst, false)	Sets the State Response to not allow override by organization
decisionSource(?resst, "AK LAW 18.13.010")	Sets the State response to reflect the decision source as State law
hasRule(?resst, 23)	Sets the rule number to 23 for reference

The corollary rule that would be invoked for any other Medical purpose is as follows:

makesRequest(?r, ?req), inState(?req, "AK"), forResource(?req, ?resource), isGeneticResult(?resource, true), forPurpose(?req, ?pur), isAKConsentRequired(?pur, true), hasResponse(?req, ?resst), responseLevel(?resst, "State"), oblName(?pre, "ConsentRequired"), clauseName(?clause, "AKGeneticConsent") -> isAllowed(?resst, true), canOverride(?resst, true), decisionSource(?resst, "AK LAW 18.13.010"), hasPreCondition(?resst, ?pre), hasClause(?resst, ?clause), hasRule(?resst, 21)

The additional arguments from the previous example are:

- **?pre** is for a specific the Pre-Condition
- **?clause** is for the text in the designated consent clause

The SWRL statements that are different than the previous rule are explained in Table III.

A side effect of these rules is that Alaska state law does not define any situation where access is denied as long as consent is obtained.

TABLE III. SAMPLE BROAD STATEMENT RULE –CONSENT REQUIRED

SWRL Statement	***Explanation***
isAKConsentRequired(?pur, true)	Restricts the rule to Purposes that do require consent
oblName(?pre, "ConsentRequired")	Gets the obligation that indicates consent is required
clauseName(?clause, "AKGeneticConsent")	Gets the specific consent clause required in Alaska for access
hasPreCondition(?resst, ?pre),	Sets the State response to include the Consent Required condition
hasClause(?resst, ?clause),	Sets the State response to include the specific Consent Clause
hasRule(?resst, 21)	Sets the rule number to 23 for reference

VI. CONCLUSION AND FUTURE WORK

Our genetic privacy ontology was built directly from the applicable Federal and State laws without any pre-conceived boundaries or required elements. The work demonstrates the importance of a purpose-focused structure to appropriately link the various data elements necessary to permit or deny access to the genetic medical information. The ontology and previous prototype work allows the data collection to be directly integrated into EHRs. The next step will be validating an integrated EHR, ontology and prototype using operational data and genetic data requests to demonstrate the appropriate data protections are enforced. This comprehensive integration reduces the provider's effort and provides access decisions in accordance with relevant laws, policies and regulations.

REFERENCES

[1] K. Wuyts, R. Scandariato, G. Verhenneman, and W. Joosen, "Integrating Patient Consent in e-Health Access Control," Int. J. Secure Softw. Eng., vol. 2, no. 2, pp. 1–24, 32 2011.

[2] J. Pritts, "The Importance and Value of Protecting the Privacy of Health Information: The Roles of the HIPAA Privacy Rule and the Common Rule in Health Research." Institute of Medicine, 2008.

[3] M. H. Ullman-Cullere and J. P. Mathew, "Emerging landscape of genomics in the electronic health record for personalized medicine," Hum. Mutat., vol. 32, no. 5, pp. 512–516, May 2011.

[4] M. D. Ritchie, E. R. Holzinger, R. Li, S. A. Pendergrass, and D. Kim, "Methods of integrating data to uncover genotype-phenotype interactions," Nat. Rev. Genet., vol. 16, no. 2, pp. 85–97, Feb. 2015.

[5] C. Pihoker et al., "Prevalence, Characteristics and Clinical Diagnosis of Maturity Onset Diabetes of the Young Due to Mutations in HNF1A, HNF4A, and Glucokinase: Results From the SEARCH for Diabetes in Youth," J. Clin. Endocrinol. Metab., vol. 98, no. 10, pp. 4055–4062, Oct. 2013.

[6] B. M. Knoppers, "Genetic information and the family: are we our brother's keeper?," Trends Biotechnol., vol. 20, no. 2, pp. 85–86, Feb. 2002.

[7] W. W. Lowrance, "Privacy, Confidentiality, and Identifiability in Genomic Research." Discussion document for workshop convened by the National Human Genome Research Institute, Bethesda, Oct-2006.

[8] M. Gymrek, A. L. McGuire, D. Golan, E. Halperin, and Y. Erlich, "Identifying Personal Genomes by Surname Inference," Science, vol. 339, no. 6117, pp. 321–324, Jan. 2013.

[9] O. for C. Rights (OCR), "Privacy Rule General Overview," HHS.gov, 05-Nov-2015. [Online]. Available: https://www.hhs.gov/hipaa/for-professionals/privacy/guidance/general-overview/index.html. [Accessed: 04-Feb-2018].

[10] O. for C. Rights (OCR), "Genetic Information," HHS.gov, 07-May-2008. [Online]. Available: https://www.hhs.gov/hipaa/for-professionals/special-topics/genetic-information/index.html. [Accessed: 04-Feb-2018].

[11] "State and Federal Consent Laws Affecting Interstate Health Information Exchange." [Online]. Available: https://www.nga.org/cms/home/nga-center-for-best-practices/center-publications/page-health-publications/col2-content/main-content-list/state-and-federal-consent-laws-a.html. [Accessed: 04-Feb-2018].

[12] "Enabling Document Sharing Using IHE Profiles - IHE Wiki." [Online]. Available: http://wiki.ihe.net/index.php/Enabling_Document_Sharing_Using_IHE_Profiles. [Accessed: 04-Feb-2018].

[13] "Basic Patient Privacy Consents - IHE Wiki." [Online]. Available: http://wiki.ihe.net/index.php/Basic_Patient_Privacy_Consents. [Accessed: 04-Feb-2018].

[14] T. G. A. for G. and Health*, "A federated ecosystem for sharing genomic, clinical data," Science, vol. 352, no. 6291, pp. 1278–1280, Jun. 2016.

[15] "Global Alliance for Genomics and Health: Privacy and Security Policy." Global Alliance for Genomics and Health, 26-May-2015.

[16] "Standards and implementation practices for protecting the privacy and security of shared genomic and clinical data." Global Alliance for Genomics and Health, 09-Aug-2016.

[17] B. M. Knoppers, "Framework for responsible sharing of genomic and health-related data," HUGO J., vol. 8, no. 1, p. 3, Dec. 2014.

[18] "HL7 Version 3 Standard: Security and Privacy Ontology, Release 1." Health Level Seven International, May-2014.

[19] H. Shen and J. Ma, "Privacy Challenges of Genomic Big Data," in Healthcare and Big Data Management, Springer, Singapore, 2017, pp. 139–148.

[20] J. M. Oliver et al., "Balancing the Risks and Benefits of Genomic Data Sharing: Genome Research Participants' Perspectives," Public Health Genomics, vol. 15, no. 2, pp. 106–114, 2012.

[21] T. Haeusermann et al., "Open sharing of genomic data: Who does it and why?," PLoS ONE, vol. 12, no. 5, p. e0177158, May 2017.

[22] H. B. Rahmouni, T. Solomonides, M. C. Mont, S. Shiu, and M. Rahmouni, "A Model-driven Privacy Compliance Decision Support for Medical Data Sharing in Europe," Methods Inf. Med., vol. 50, no. 4, pp. 326–336, 2011.

[23] H. B. Rahmouni, T. Solomonides, M. C. Mont, and S. Shiu, "Privacy compliance and enforcement on European healthgrids: an approach through ontology," Philos. Trans. R. Soc. Lond. Math. Phys. Eng. Sci., vol. 368, no. 1926, pp. 4057–4072, Sep. 2010.

[24] B. Blobel, "Ontology driven health information systems architectures enable pHealth for empowered patients," Int. J. Med. Inf., vol. 80, no. 2, pp. e17–e25, Feb. 2011.

[25] F. Falcao-Reis, A. Costa-Pereira, and M. Correia, "Access and privacy rights using web security standards to increase patient empowerment," in Medical and Care Compunetics 5, vol. 137, IOS Press, 2008, pp. 275–285.

[26] P. Hung and Y. Zheng, "Privacy Access Control Model for Aggregated e-Health Services," EDOC Conf. Workshop 2007 EDOC 07 Elev. Int. IEEE, no. 15–16 Oct. 2007, pp. 12–19, Jul. 2008.

[27] Genetic Information Nondiscrimination Act of 2008. 2008.

[28] "Genome Statute and Legislation Database," National Human Genome Research Institute (NHGRI). [Online]. Available: https://www.genome.gov/. [Accessed: 04-Feb-2018].

[29] Health Insurance Portability and Accountability Act of 1996. 1996.

[30] W. M. P. van der Aalst and A. H. M. ter Hofstede, "YAWL: yet another workflow language," Inf. Syst., vol. 30, no. 4, pp. 245–275, Jun. 2005.

[31] H. Knublauch, R. W. Fergerson, N. F. Noy, and M. A. Musen, "The Protégé OWL Plugin: An Open Development Environment for Semantic Web Applications," in The Semantic Web – ISWC 2004, 2004, pp. 229–243.

[32] M. Reep, B. Yu, D. Wijesekera, and P. Costa, "Sharing Data under Genetic Privacy Laws," in Proceedings of the Eleventh Conference on Semantic Technology for Intelligence, Defense, and Security, Fairfax VA, USA, 2016, pp. 46–54.

[33] M. Reep, B. Yu, D. Wijesekera, and P. Costa, "Sharing Genetic Data under US Privacy Laws," in 11th International Joint Conference on Biomedical Engineering Systems and Technologies, Funchal, Madeira, Portugal, 2018, vol. 5: HEALTHINF, pp. 349–360.

Fiber Bragg Grating Sensors for Temperature Monitoring during Radio Frequency Thermal Ablation (RFTA) Treatment on *Ex-Vivo* Organs

Giovanna Palumbo, Agostino Iadicicco, and Stefania Campopiano
Dept. of Engineering, University of Naples "Parthenope"
Naples, Italy
email: giovanna.palumbo@uniparthenope.it, iadicicco@uniparthenope.it, campopiano@uniparthenope.it

Paolo Verze, Nicola Carlomagno, Vincenzo Tammaro, and Juliet Ippolito
Dept. of Urology and Dept. of General Surgery
University of Naples "Federico II"
Naples, Italy
email: p.verze@gmail.com, nicola.anita@tiscali.it, vincenzo.tammaro@unina.it, juliet@unina.it

Paola Saccomandi
Institute of Image-Guided Surgery of Strasbourg
Strasbourg, France
email: p.saccomandi@unicampus.it

Elena De Vita, Carlo Massaroni and Emiliano Schena
Università Campus Bio-Medico di Roma,
Rome, Italy
email: devita.ele@gmail.com, c.massaroni@unicampus.it, e.schena@unicampus.it

Daniele Tosi
School of Engineering, Nazarbayev University,
Astana, Republic of Kazakhstan
email: daniele.tosi@nu.edu.kz

***Abstract*—In this work, we report on the use of fiber Bragg gratings (FBGs) sensor for real-time in-situ temperature monitoring during Radio Frequency Thermal Ablation (RFTA) tumor treatment. In order to create a thermal multi-point measurement of the area to be treated, a proper sensing configuration was developed by instrumenting the RF commercial probe with more than 30 FBG sensors. The experiments were conducted on ex-vivo animal liver and kidney tissues and results confirm that the proposed setup is able to conduct a multi-point measurement and to distinguish between different and consecutive RF discharges with a temperature resolution of 0.1 °C and a minimum spatial resolution of 1 mm.**

Keywords-Fiber Bragg Gratings (FBGs); Radio Frequency Thermal Ablation (RFTA); temperature monitoring.

I. INTRODUCTION

RFTA is a local treatment of small tumors that, by using RF current released by electrodes connected to a RF generator, locally induces a rapid temperature increase in the tissue affected by the tumor with resulting immediate necrosis of diseased tissues [1]. Real-time temperature monitoring during the treatment is essential in enabling surgeons to adapt RFTA parameters to the tumor size during surgery and consequently preserve as much healthy tissue as possible. Due to their small size, non-toxicity, chemical inertness, biocompatibility, flexibility and low cost, Fiber Bragg Gratings seems to be a good solution to monitor the temperature during the RF treatments [2].

In Section II, a novel setup to temperature monitoring during RFTA treatment is proposed and the results of a RF discharge on ex vivo animal liver are shown, while in Section III the conclusion is reported.

II. EXPERIMENTAL SETUP AND RESULTS

The experimental setup is illustrated in Figure 1. In particular, in the blue box, the RFTA instrumentation is reported: the RF probe is powered by a 500 kHz generator (RITA 1500X®) producing up to 250 W of power. The laparoscopic bipolar RF device Habib 4x® was used, consisting of two pair of opposing electrodes with active ends of 6 cm in length. In the red box, it is also reported the optical interrogation of the FBGs' sensor used to measure the reflected signal. This device consists of a commercial FBG interrogator in the range of 1500-1600 nm with a resolution of 1 pm and maximum sampling frequency of 1 Sample/s. Several commercial FBGs 1 mm long have been used. In order to correctly measure the temperature during the RF discharges, the FBGs were inserted in carbon fiber needles and properly fixed to the RF probe, as shown in Figure 2. Several measurements of the spatial temperature profile during multi-step RFTA discharges and the monitoring of the heat propagation on animal liver and kidney organs have been carried out confirming that the proposed setup is able to conduct a multi-point measurement and to distinguish between different and consecutive RF discharges with a temperature resolution of 0.1 °C and a minimum spatial resolution of 1 mm. For example, in Figure 3, the temperature is reported by a color map versus spatial sensor position along the axis parallel and perpendicular to the electrodes respectively (vertical axis in the graph) and time (horizontal axis).

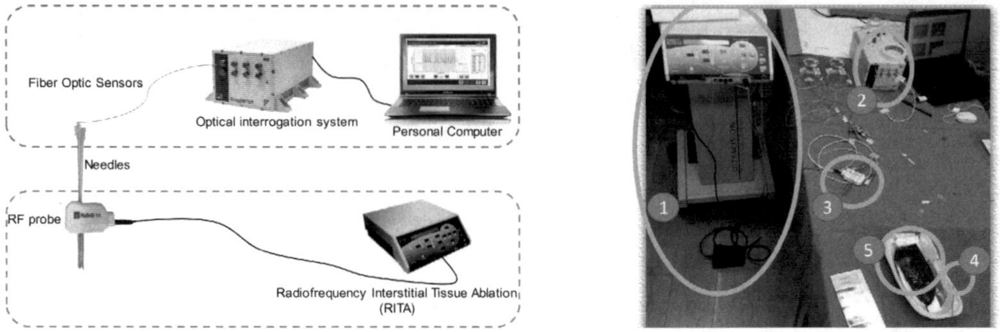

Figure 1. Left: Schematic of the experimental measurement setup. Right: Picture of the experimental setup where 1) the RF generator RITA; 2) the optical interrogation system FS22; 3) the modified RF probe; 4) thermocouple used as a reference; 5) the organ sample.

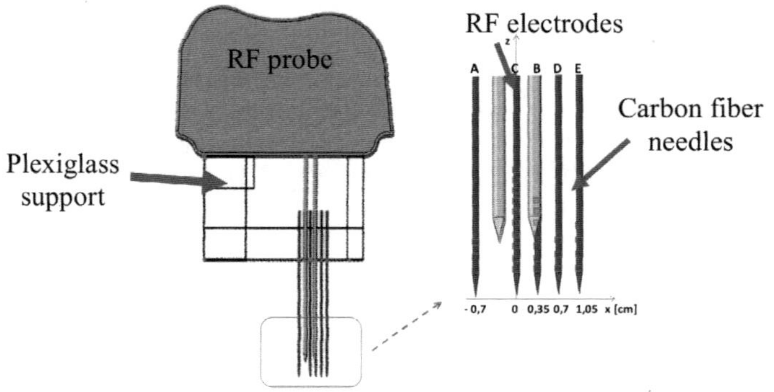

Figure 2. Schematic of the RF probe instrumented with FBGs inserted in carbon fiber needles fixed to the probe.

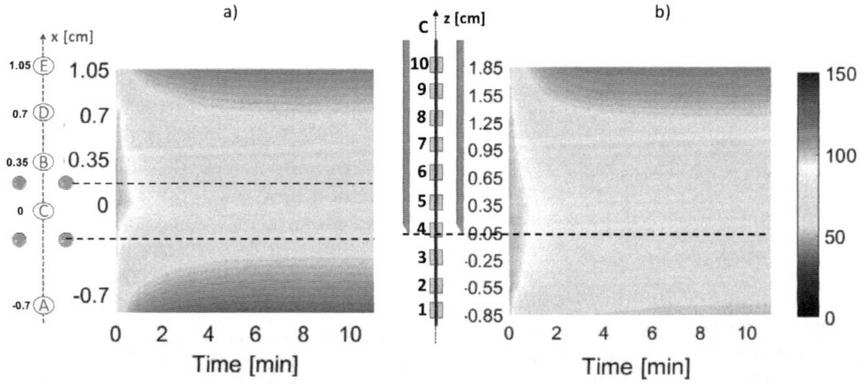

Figure 3. (a) Temperature profile of the FBGs at the end of electrodes. (b) Temperature profile of the 10 FBG sensors of the array C tip.

III. Conclusion

Thanks to the FBG's fast response time and small size, we were able to monitor temperature with a resolution of 0.1 °C during RF discharges. This mapping of the tissue temperature allowed the monitoring of the thermal dose delivered to the patient in real time and, at the same time, to avoid potentially adverse effects to the tissue adjacent to the target region.

References

[1] K. F. Chu and D. E. Dupuy, "Thermal ablation of tumours: biological mechanisms and advances in therapy," Nat. Rev. Cancer, vol. 14, no. 3, pp. 199–208, Feb. 2014.

[2] G. Palumbo et al., "Temperature profile of ex-vivo organs during radio frequency thermal ablation by fiber Bragg gratings," J. Biomed. Opt., vol. 21, no. 11, p. 117003, 2016.

Virtual Coach: Predict Physical Activity Using a Machine Learning Approach

Talko Dijkhuis[1,2], Johan Blok[1] and Hugo Velthuijsen[1]
[1]Hanze University of Applied Sciences, Institute for Communication, Media & IT, Groningen, The Netherlands
[2]University of Groningen, Johann Bernoulli Institute for Mathematics and Computer Science, Groningen, The Netherlands
[1]{t.b.dijkhuis, j.blok, h.velthuijsen}@pl.hanze.nl

Abstract—**One of the main causes of numerous health problems is a lack of physical activity. To promote a more active lifestyle, the Hanze University started a health promotion program. Participants were motivated to reach their daily goal of physical activity by means of an activity tracker in combination with two-weekly coaching sessions. Employing the data of the experiment, we investigated the manners in which the predictability of physical activity of a participant during the day can be improved. The collected step count data was used to construct personalised machine learning models, by taking into account the difference between physical activities during weekdays on the one hand and weekends on the other hand. The training of algorithms per participant in combination with the time-slices weekdays, weekend and the whole week improves the accuracy of the prediction model. The performance of the models improves even further when the individualised time-sliced models are combined. More contextual data, like free time and working hours, might even extend the accuracy. The use of personalised prediction models, based on machine learning and time slices, could become an addition in preventive personalized eHealth systems and mobile activity monitoring. For instance, this can constitute as a viable addition to a virtual coaching system to help the participants to reach their daily goal. As the individualised models allow for predictions of the progression of the physical activity during the day, they enable the virtual coaching system to intervene at the appropriate moment in time.**

Keywords—**preventive eHealth systems; monitoring physical activity; machine learning; prediction; virtual coach.**

I. INTRODUCTION

An unhealthy lifestyle with insufficient daily physical activity shortens life expectancy. Not meeting the recommended level of physical activity is associated with 5.3 million deceases globally in 2008 [1]. Lack of physical activity is also associated with a decreased quality of life, lower levels of social participation, and disability to work. In the workplace employees with low and medium physical activity have a 2.4-3.5 fold higher rate of unplanned illness-related absenteeism compared to people who meet the Centers for Disease Control and Prevention (CDC) guideline of 150 min/week [2].

The negative effects of lacking physical activity have fostered a novel initiative at the Dutch University, Hanze University of Applied Sciences (HUAS). The university started an initiative to promote a healthy lifestyle and physical activity during the workday called (in Dutch): Het Nieuwe Gezonde Werken (The New Healthy Way of Working; HNGW). This initiative on promotion of a healthy lifestyle included a focus on the improvement of physical activity. Participants got an activity tracker to increase the awareness of their daily progress in achieving their goals in terms of numbers of steps. The daily feedback of the activity tracker was complemented with a fortnightly coaching session on the lifestyle and the physical activity. However, the feedback of the activity tracker and its platform didn't provide the participant with timely personalised feedback. Neither was the coach timely informed with information on the participant to enable a personalised intervention. Furthermore, current activity trackers do not provide a probability of reaching the daily goal or take the difference between weekdays and weekend into account, although this difference is known for a different level of activity [3].

In this paper, we propose a personalised, flexible machine learning based model that enables personalized eHealth being supported by preventive systems on activity tracking. The personalised model enables feedback on a participant's probability of reaching his or her daily activity goal. The first section introduces the state of the art on measuring activity levels, the use of machine leaning and monitoring. Subsequently, we describe the study on health promotion at HUAS, the collected dataset on daily physical activity of the participants, the method of statistical analysis of the results trained algorithms, and the selection and training of the algorithms. In the third section, we present the results of the training of the algorithms and the statistical analysis. The conclusion on the results and a short discussion on future work finish this paper.

II. STATE OF THE ART

Activity trackers provide a measure for the number of steps humans make and enable monitoring. Adding a step counter to physical therapy or counselling was effective in some groups [4] [5]. The collection of step data is not only effective for therapy or counselling, it is also an intervention mechanism in itself [6]. Only the fact of using an activity tracker could motivate physical activity and improvement of health [7]. To improve on physical activity in combination with activity tracking monitoring, coaching is helpful. Perceiving the information personal and in context and timely is important for the effectiveness of (e)Coaching [8]. The participant needs to receive the information and the advice while it is relevant. To the best of our knowledge no studies exist on the use of activity trackers in combination with machine learning algorithms to establish *individualized* models or studies on *individualized* models used in preventive systems on monitoring activity helping the participant to improve his or her physical behaviour.

III. METHODS

In this section, we present the study design of the HNGW, the data set we used to train the algorithms and the methods used for statistical analysis.

A. Study design

The study data stems from the HNGW project. Forty-eight healthy employees were recruited from the HUAS. The 48 participants were divided according to age, gender, BMI, and baseline self-reported health prior to being randomized into two groups. Group A followed a twelve-week health promotion intervention; the other group, group B, served as a control first and thereafter received the intervention. The outcome measures included, among other values, the daily steps. The daily steps were measured with the Fitbit Flex, which is known to be a trustworthy and valid activity tracker for step count and suitable for health promotion programs [9].

1) Data set: In order to prepare the available minute step data as input for training the algorithms, we followed a step-by-step approach. First, we performed a data pre-processing step to remove the incomplete records from the data set. We also eliminated all records per day whenever no step was gathered during that day. Second, we constructed an hourly summarised data set with several new derived variables representing:

1) the day of the week (range 0 - 6)
2) hour of day (range 0 - 23)
3) week of the year (range 0-52)
4) year (2014-2015)
5) a cumulative sum of the steps per hour

Third, a workday is defined as the weekdays Monday till Friday. The normal working hours at the university are between 08:00AM and 05:00PM. The project tried to motivate the participants to walk at least a part of the distance they daily commute. As a consequence, the hours of interest are the combination of the working hours and commuting. Therefore, we sliced the dataset such that it only contains the number of steps per hour, per workday between 07:00AM and 06:00PM. Fourth, a weekend is defined as Saturday and Sunday. In order to enable comparison with the weekdays, we sliced the data set for the weekend in the same way as the workdays, per weekend day between 07:00AM and 06:00PM. Fifth, partial sum of steps per hour throughout the day was included. Sixth, a column was added comprising the average number of steps at 06:00PM calculated over all weeks. For this average the amount of steps between 7:00AM and 6:00PM was considered. This column was regarded as a threshold in order to determine the outcome column. Finally, we constructed a binary outcome variable based on the threshold.

2) Statistical Analysis: Four different algorithms were trained. To compare the performance of the algorithms, we used the confusion matrix method to classify the difference between the predicted value and the actual value. A confusion matrix provides an overview of the true positives (TP; a predicted a 'true' and the actual data contained a 'true'), true negatives (TN; the model predicted a 'false' and the actual data was a 'false'), false positives (FP; the model predicted a 'true' label, but the actual data was a 'false'), and false negatives (FN; the model predicted a 'false' label, but the data was 'true') of a model. The confusion matrix served as a basis for the calculation of the performance measure F1-score [10].

The F1-score was calculated for each model, the F1-score has a range of zero to one, one is the best score. To calculate the F1-score, two other metrics known as the precision and the recall are used. Precision is the proportion of the true positives and the false negatives, and is calculated as $\frac{TP}{(TP+FN)}$. Recall is the true positive rate, which is calculated as $\frac{TP}{(TP+FP)}$. Using precision and recall, the F1-score is calculated as $2 \cdot \frac{Precision \cdot Recall}{Precision+Recall}$.

B. Selection of algorithms

The goal is to predict, during the day, whether a participant will reach his daily number of steps. This is known as a classification problem. The selection of the best algorithm is a matter of trial and error. It is generally agreed upon that it's not possible to determine the best performing algorithm upfront [11]. The general approach for solving this problem is very similar to the travelling salesman problem [12]. Although there are classes of algorithms which are more suitable for different types of problems. One of the biggest open-source community on machine learning, scikit-learn.org provides a 'flowchart' with rough indications which algorithms may perform best [13]. We choose four possibly well-performing algorithms: (i) ADAboost (ADA), (ii) Decision Tree (DT), (iii) Random Forest (RF), and (iv) Stochastic Gradient Descent (SGD). After splitting the data in a training- and a test set, the performance of each of the algorithms was calculated.

Firstly, based on the whole training set the F1-score of each algorithm was determined. Secondly, the algorithms were trained on the individualized training data utilizing three different time slices of the dataset and the trained algorithms were converted into to individualized time slice based models (TSM):

TSM1:work week (Monday-Friday)
TSM2:weekend (Saturday, Sunday)
TSM3:whole week (Monday-Sunday)

The result of the training was of 12 different models per participant (TSM 1-3 times the four algorithms). Next the ranking and the overall best performing algorithm was determined.

Thirdly, for the three personalized time sliced models of the overall best performing algorithm, the F1-score was calculated using the complete data set. TSM 1 was used to calculate the

F1-score for the workweek, TSM 2 was used to calculate the F1-score for the weekend, and TSM 3 was used to calculate two F1-scores, respectively for the work week and for the weekend.

Fourthly, the combination of week F1-score and weekend F1-score of the diverse time slice models were studied on the performance.

IV. RESULTS

The group F1-scores per algorithm were for Random Forest 0.89, Decision Tree 0.88, ADAboost 0.69, and Stochastic Gradient Descent 0.44. Application of the individualized component and time slices slightly improved the performance. Only ADAboost showed big differences on the F1-score. Figure 1 displays the results of the average of the individual scores on the subsequent algorithms and time-slices.

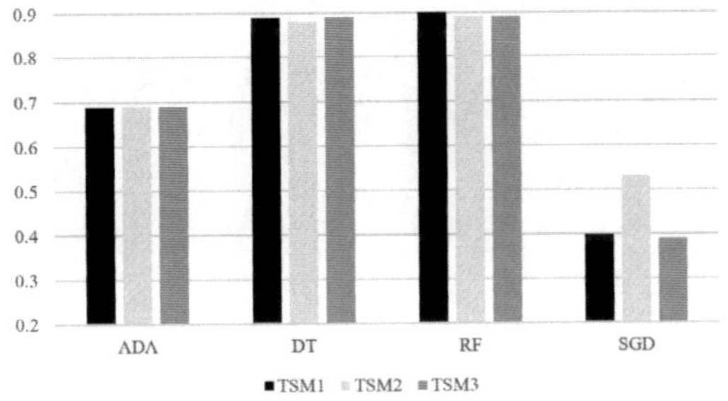

Fig. 1. Average F1-score of the time-sliced models over all participants.

Table I represents the numbers of the average of the individual scores.

TABLE I. AVERAGE F1-SCORE OVER ALL PARTICIPANTS OF THE TIME-SLICED MODELS.

	ADA	*DT*	*RF*	*SDG*
TSM1	0.69	0.89	0.9	0.4
TSM2	0.69	0.88	0.89	0.53
TSM3	0.69	0.89	0.89	0.39

On a group level the best performing algorithm is Random Forest. The Random Forest based, individual time sliced models in different combinations resulted in diverse best combinations. Table II states the diverse combinations of the time slice models and the F1-score. For 29 participants there is one ideal combination, for 12 participants there are 2 best combinations and for 3 participants the diverse combinations of the time slice models perform equally. All combinations of different individualised time slice models outperformed the group result stated in Table II.

TABLE II. COMBINATIONS OF TIME-SLICES AND THEIR F1-SCORE.

	A & B	*A & D*	*C & B*	*C & D*
one best	6	11	9	3
F1-score	0.95	0.95	0.92	0.93
standard deviation	0.03	0.02	0.03	0.02
two equally best	6	5	7	6
F1-score	0.95	0.95	0.95	0.95
standard deviation	0.03	0.02	0.03	0.02
all equal	3	3	3	3
F1-score	0.96	0.96	0.96	0.96
standard deviation	0.02	0.02	0.02	0.02

A: TSM1, work week (range: Monday-Friday)
B: TSM2, weekend (range: Saturday, Sunday)
C: TSM3, work week (range: Monday-Friday)
D: TSM3, weekend (range: Saturday, Sunday)

V. CONCLUSION

The individualisation of the machine learning models improved the F1-score in comparison to the group level F1-score. The best performing algorithm was the Random Forest algorithm. Application of the literature based thesis concerning the difference of physical activity between week days and weekend for training different algorithms improved the F1-score. It is recommended to construct time sliced weekend and week models per individual and calculate which combination of models performs best. To improve the performance of the individualised models in the future, contextual data that influences physical activity, like free time, regular physical activity, and illness, may be taken into account. The individualisation of the predictive models enables automated personalised timely coaching. The results of this paper will be applied in the preventive eHealth virtual coach platform as suggested by Blok et al. [14]. A possible future direction is to create a model per day per individual.

ACKNOWLEDGMENT

The authors thank the Hanze University Health Program, especially M. van Ittersum, for providing the physical activity data of the Health Program, all the participants in the experiment, and F. Blaauw for suggestions on the application of machine learning.

REFERENCES

[1] I. Min-Lee *et al.*, "Effect of physical inactivity on major non-communicable diseases worldwide: An analysis of burden of disease and life expectancy," *The Lancet*, vol. 380, no. 9838, pp. 219–229, 2012. [Online]. Available: http://dx.doi.org/10.1016/S0140-6736(12)61031-9

[2] E. Losina, H. Y. Yang, B. R. Deshpande, J. N. Katz, and J. E. Collins, "Physical activity and unplanned illness-related work absenteeism: Data from an employee wellness program," *PLoS ONE*, vol. 12, no. 5, pp. 1–8, 2017.

[3] C. E. Matthews *et al.*, "Sources of Variance in Daily Physical Activity Levels in the Seasonal Variation of Blood Cholesterol Study," *American Journal of Epidemiology*, vol. 153, no. 10, pp. 987–995, 2001.

[4] H. J. de Vries, T. J. Kooiman, M. W. van Ittersum, M. van Brussel, and M. de Groot, "Do activity monitors increase physical activity in adults with overweight or obesity? A systematic review and meta-analysis," *Obesity*, vol. 24, no. 10, pp. 2078–2091, oct 2016. [Online]. Available: http://doi.wiley.com/10.1002/oby.21619

[5] M. Miyauchi *et al.*, "Exercise Therapy for Management of Type 2 Diabetes Mellitus: Superior Efficacy of Activity Monitors over Pedometers," *Journal of Diabetes Research*, vol. 2016, pp. 1–7, sep 2016. [Online]. Available: https://www.hindawi.com/journals/jdr/2016/5043964/

[6] L. A. Cadmus-Bertram, B. H. Marcus, R. E. Patterson, B. A. Parker, and B. L. Morey, "Randomized Trial of a Fitbit-Based Physical Activity Intervention for Women." *American journal of preventive medicine*, vol. 49, no. 3, pp. 414–8, sep 2015. [Online]. Available: http://www.ncbi.nlm.nih.gov/pubmed/26071863http://www.pubmedcentral.nih.gov/articlerender.fcgi?artid=PMC4993151

[7] Z. H. Lewis, E. J. Lyons, J. M. Jarvis, and J. Baillargeon, "Using an electronic activity monitor system as an intervention modality: A systematic review," *BMC Public Health*, vol. 15, no. 1, p. 585, dec 2015. [Online]. Available: http://bmcpublichealth.biomedcentral.com/articles/10.1186/s12889-015-1947-3

[8] M. Gerdes, S. Martinez, and D. Tjondronegoro, "Conceptualization of a Personalized eCoach for Wellness Promotion," in *Proceedings of ACM 11th EAI International Conference on Pervasive Computing*. Barcelona: Association for Computing Machinery, 2017.

[9] J. Wang, R. Chen, X. Sun, M. F. H. She, and Y. Wu, "Recognizing human daily activities from accelerometer signal," *Procedia Engineering*, vol. 15, pp. 1780–1786, 2011, URL: http://dx.doi.org/10.1016/j.proeng.2011.08.331 [accessed:2018-01-31].

[10] S. Schoeppe *et al.*, "Efficacy of interventions that use apps to improve diet, physical activity and sedentary behaviour: a systematic review," *International Journal of Behavioral Nutrition and Physical Activity*, vol. 13, no. 1, p. 127, dec 2016, URL: http://ijbnpa.biomedcentral.com/articles/10.1186/s12966-016-0454-y [accessed:2018-01-31].

[11] D. H. Wolpert, "The lack of a priori distinctions between learning algorithms," *Neural Computation*, vol. 8, no. 7, pp. 1341–1390, 1996, URL: http://www.mitpressjournals.org/doi/10.1162/neco.1996.8.7.1341 [accessed:2018-01-31].

[12] S. Raschka and V. Mirjalili, *Python Machine Learning*. Packt Publishing Ltd, 2015.

[13] "scikit learn,choosing the right estimator," URL: http://scikit-learn.org/stable/tutorial/machine_learning_map/index.html [accessed:2018-01-31].

[14] J. Blok, A. Dol, and T. Dijkhuis, "Toward a Generic Personalized Virtual Coach for Self-management : a Proposal for an Architecture," no. c, pp. 105–108, 2017.

Exploring the Use of Context-Awareness in Scheduling Methods to Approach the Patient Planning Problem

Conceição Granja and Terje Solvoll
Norwegian Centre for e-Health Research
University Hospital of North Norway
Tromsø, Norway
e-mail: {conceicao.granja; terje.solvoll}@ehealthresearch.no

Abstract—**Lately, in literature, a few papers have been published on methods for solving patient planning problems involving uncertainty. Most authors consider very simple recourse actions, typically only counting the extra overtime resulting from unexpected delays. This is not very realistic in a real world setting, since one in practice wants to dynamically reschedule as unexpected events occur. To be able to perform as such, a scheduling system must address three main challenges: (1) enlarge its scope; (2) plan for uncertainty; and (3) solutions approach. In this paper, is discussed how the context-aware methodology can contribute as a solution to this problem, by enabling process support.**

Keywords-scheduling; context-aware; healthcare; workflow

I. INTRODUCTION

The majority of the hospitals use a so-called block-booking system when planning surgeries. In this system, a medical specialty is assigned to blocks denoting a specific amount of time, e.g., a day, in one Operating Room (OR). These blocks can be combined into cyclical Master Surgery Schedules (MSS), where every block is repeated after a fixed cycle. At the strategic level of block-booking system, the number of blocks assigned to the specialties and emergencies during a MSS cycle is determined. At the tactical level, OR-days are allocated to specialties in an MSS, such that the strategic allocation is met.

In recent years, approaches to solve MSS have become more complex, in the sense that they started to consider multiple resources and tackle uncertainty more accurately.

Uncertainties might come from different sources, such as processing times, demand/patient arrivals, no-show ups, personnel availability, etc.

Clearly, MSS affects the patient flow to downstream inpatient care units. Surgeries performed in each block of the MSS create a flow of patients through the Intensive Care Unit (ICU) to the ward, or directly from the OR to the wards, before they leave the hospital. The post-anesthesia care unit might be part or not of the OR department.

The development of an MSS module must address three main challenges:

1) Enlarging the scope of the MSS: MSS approaches embedded in commercial software consider only the impact of the MSS on operating theatre and operating staff; the goal here is to enlarge the scope to down-stream resources, such as the intensive care unit ICU and the general wards required by the patients. The solution module should be flexible enough to cope with different features that appear in different hospitals that interfere with the planning activities.

2) Planning with uncertainty: Surgical management processes are subject to high variability resulting in significant deviations between intended and actual performance of surgical plans. For instance, when surgeries take longer than predicted or emergency patients arrive, it often results in overtime and possible cancellation of surgeries. When planning at an aggregate level, uncertainties are usually neglected. The challenge is to anticipate the uncertainties and incorporate them during the MSS decision-making.

3) Solution approaches: The problem cannot be totally described in mathematical programming terms. The volatility of information (see previous point) makes it difficult to incorporate all uncertainty in a single solid deterministic model.

To tackle such challenges, a MMS module has to enable a fast and automated, fully context dependent, scheduling. In such scenario, context-aware systems present themselves as a promising approach.

This paper is divided in four section. In Section I, are presented the major challenges regarding scheduling in healthcare environments. In Section II, is presented a brief literature review on evidence that some of the health IT, currently implemented in clinical practice, lacks process support. In Section III, is presented the context-aware methodology, and, in Section IV, is discussed how this methodology can contribute to process support, and improvement of operational management. Section V concludes the article providing some ideas for future work.

II. BACKGROUND

A few papers have lately been published on methods for solving patient planning problems involving uncertainty. Significant improvements can be attained by doing this [1]-[4]. Exact methods, such as stochastic programming [1][2], or robust optimization [2] seem unable to solve problems of a realistic size within reasonable time. For this reason, some authors apply a "Sample Average Approximation" (SAA) approach, to speed up the calculation of recourse costs (i.e., the cost of handling unforeseen changes in the scheduling problem) [3][5]-[7]. Even more pragmatic approaches use

Column generation [6], simple heuristics [2], local search [8], or meta-heuristics like Simulated Annealing [5][8] or Tabu search [5]. These can be applied either to the original problem, to the SAA simplified problem, or to a modified problem including hedging on resource demand or availability [8].

Most authors consider very simple recourse actions, typically only counting the extra overtime resulting from unexpected delays. This is not very realistic in a real world setting, since one in practice wants to dynamically re-schedule as unexpected events occur. Only very few authors consider more advanced recourse actions [4][9].

Very little work has been done for integrated planning and scheduling under uncertainty for planning that covers more than a week into the future. The vast majority of hospitals do not consider data uncertainty at the planning level. However, such an approach can have substantial negative impact at the operational level and result in a suboptimal use of OR's.

For obtaining robust schedules, that will anticipate possible disruptions, OR's planning should consider uncertainty. Bruni et al. [4] use stochastic programming to model uncertainty associated with arrival of emergency patients and duration of surgery. The authors also presented some recourse strategies that model reactive scheduling policies.

The problem of optimizing the assignment of surgeries and sufficient planned slacks to the operating days such that the risk of working overtime is minimized, no surgeries are cancelled and operating room utilization is improved is addressed by [8]. The problem is solved by Local Search and improvement heuristics.

Other models, that create MSSs with levelled bed occupancy in downstream units, are presented in [10][11]. Both the number of patients and the length of stay in the hospital are assumed to be stochastic. Beliën and Demeulemeester [10] aim to minimize the expected bed shortage. The problem is solved by mixed-integer programming and Simulated Annealing. Fügener et al. [11], concentrate on inpatient flow and define a model to calculate, for a given MSS, the expected distributions of recovering patients in the downstream units. Based on this, it is proposed an approach for planning the MSS with the objective to minimize downstream costs by levelling bed demand and reducing weekend bed requests. The main distinction between the two papers is that [10] only allows one downstream resource, while [11] models multiple downstream units.

A multistage stochastic mixed-integer programming formulation for the assignment of surgeries to operating rooms over a finite planning horizon is proposed in [12]. Demand for and the duration of surgery are random variables and the objective is to minimize expected cost of surgery cancellations, patient waiting time, and operating room overtime.

It should be emphasized again, that the master surgery schedules are usually performed manually at the hospitals, without any type of systems to support the decision making. It should be performed with a given periodicity. The underlying method to generate MSS that levels the workload and increases the efficiency of the surgical nursing wards, without deteriorating the OR department's efficiency, is based on simulation-optimization.

The above-described work identifies common signs that the implemented technology lacks process support. To complete care processes, health personnel work as a team, performing high risk tasks under uncertainty, and time pressure, dependent on a wide and reliable communication infrastructure for exchanging different kinds of data, such as patient reports, lab tests and working shifts, together with text, voice and alarm services. The management of this information is difficult and requires considering a wide variety of problems that should be avoided in order to properly meet the needs of hospital professionals. Context-awareness can provide the necessary knowledge for health IT to reduce inefficiencies and manage complexity.

III. Materials and Methods

Let us start by defining "context". To define "context", some of the definitions given by the research community [13]-[17] over the years were investigated, and it is concluded that the most suitable definition for our research is [18]:

"Context is any information that can be used to characterize the situation of an entity. An entity is a person, place, or object that is considered relevant for the interaction between a user and an application, including the user and applications themselves."

This definition shows the importance of which information is relevant or not in a context-sensitive system. A context-sensitive system could, therefore, be defined as a system allowing interactions between multiple entities using relevant information. Abowd et al. [18], state that: "*A system is context-aware if it uses context to provide relevant information and/or services to the user, where relevancy depends on the user's task*". This definition shows that a context-sensitive system can change its behavior and send some relevant information according to the context, which reflects our view.

The trend in the health IT field has been to push as much information as possible to the users, in order to provide more sophisticated and useful services while, at the same time, making users more available. During a preliminary research study on the AwareMedia system [19], a classification that splits the above listed information along three main axes is suggested:

- Social awareness: `where a person is', `activity in which a person is engaged in', `self-reported status';

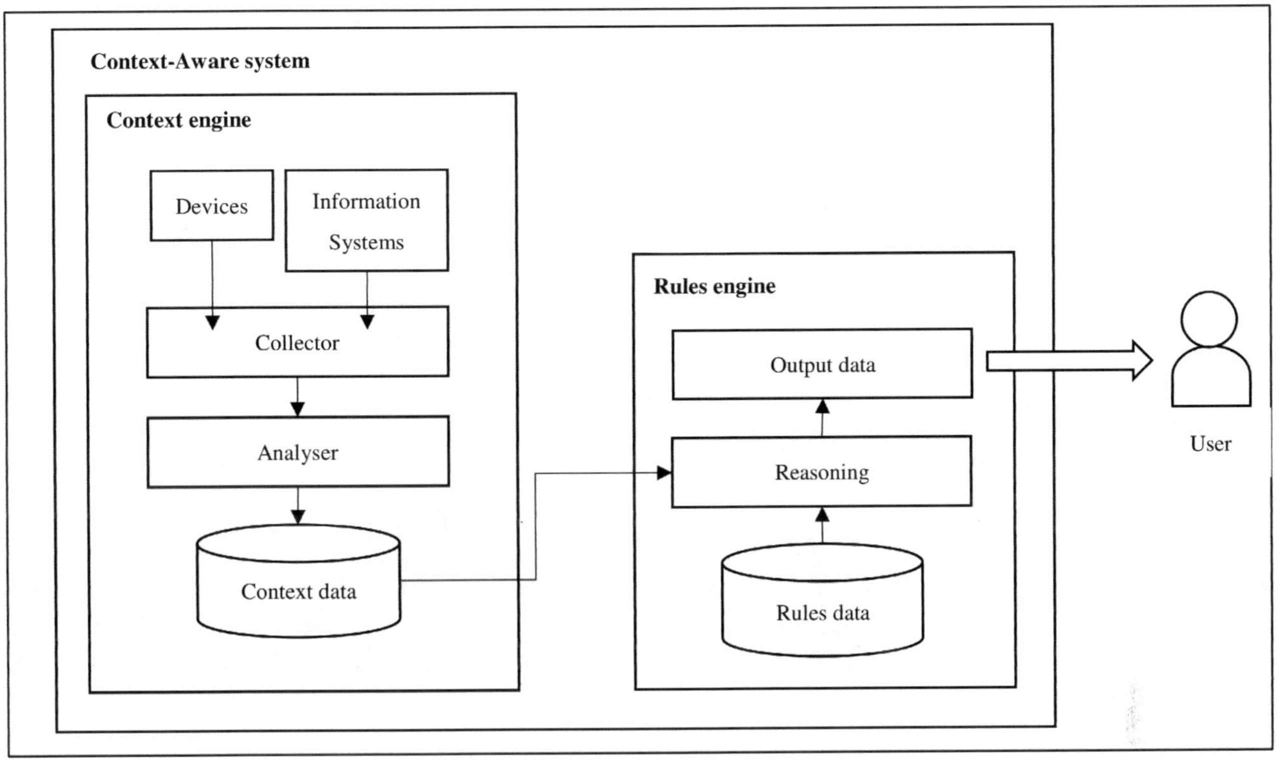

Figure 1. Illustration of context-aware systems' basic architecture.

- Spatial awareness: 'what kind of operation is taking place in a ward', 'level of activity', 'status of operation and people present in the room';
- Temporal awareness: 'past activities', 'present and future activities' that is significant for a person.

A context-aware system, as shown in Figure 1, comprises two main modules:

- Context engine: This module interfaces with other information systems and devices to collect raw data. These are then fed to an analyzer to classify raw data and generate context data.
- Rules engine: This module acts as filter between the data and the user. By applying a set of pre-defined conditions that define what, when, and to who the information must be presented. Such rules can be defined manual or automatically.

The adoption of context-aware systems based on these definitions is growing in a variety of domains, such as smart homes, airports, travel/entertainment/shopping, museum, and offices, as mentioned in [20].

IV. Discussion and Conclusions

Health IT usability, and adoption in daily practice is closely related to the systems' semantic and technological interoperability. This requires the systems to provide a comprehensive platform for process support. On the other hand, to provide this platform, structured knowledge that is not currently available in the Electronic Health Record (EHR) systems in use in most Norwegian hospitals would be required.

V. Future Work

Technological interoperability can be achieved by describing clinical guidelines using standardized languages. The context-aware methodology described Section III can support both the knowledge and technological interoperability required.

A context-aware system can collect data not only from the EHR, but also from the other IT implemented at the hospital. Such data can then be made available in different patient settings, and processed according to rules, to generate new knowledge. A context-aware system can also learn from the user interaction with the system to automatically improve his/her experience. In this manner, a context-aware system is able to provide process support by analyzing process related data of two categories: (1) "what is done"; (2) "how it is done".

The progression of a patient in a clinical process is determined by the completion of the tasks that compose the same process. However, EHR systems are not always updated on the tasks' completion as different individuals evidence different work patterns. If technology is able to separate the process related data as described above, then it becomes possible to achieve adaptive workflows.

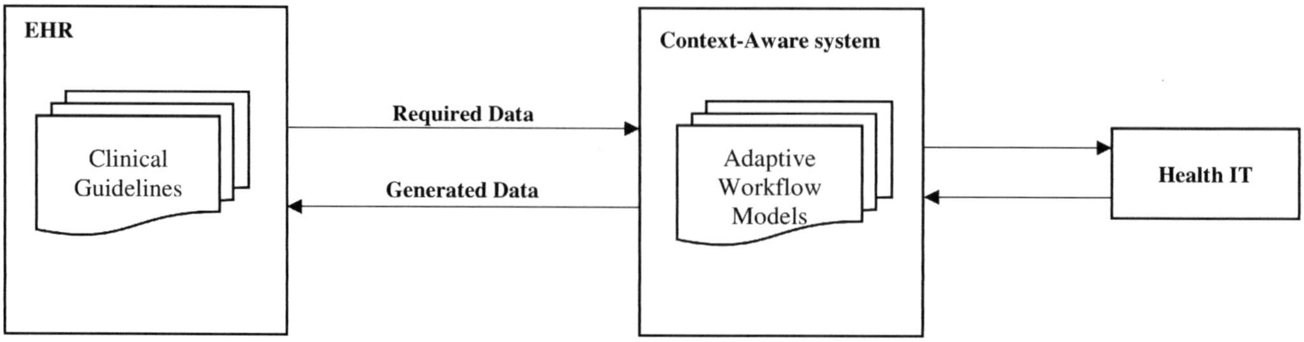

Figure 2. Illustration of the proposed context-aware based health IT system architecture.

"What is done" can be described by translating clinical guidelines using a standardize language like OpenEHR archetypes.

"How it is done" can be achieved by using machine-learning techniques, fed with context data, to adjust the clinical guideline to the individual user work pattern. The semantic interoperability is achieve through the definition of the data required to support workflow on the individual level to bring both concepts together using OpenEHR archetypes. An illustration of the system architecture is presented in Figure 2.

Context-awareness allows health IT to provide process support by managing the complexity inherent to clinical processes while supplying the technology with the process standards required to ensure usability.

ACKNOWLEDGMENTS

The authors would like to thank the regional health authority Helse-Nord for funding the research project HST 1304-16.

REFERENCES

[1] S. Batun, B. T. Denton, T. R. Huschka, and A. J. Schaefer, "Operating Room Pooling and Parallel Surgery Processing under Uncertainty," INFORMS journal on Computing, vol. 23, pp. 220-237, 2011.

[2] B. T. Denton, A. J. Miller, H. J. Balasubramanian, and T. R. Huschka, "Optimal Allocation of Surgery Blocks to Operating Rooms under Uncertainty," Operations research, vol. 58, pp. 802-816, 2010.

[3] M. Lamiri, X. Xie, A. Dolgui, and F. Grimaud, "A Stochastic Model for Operating Room Planning with Elective and Emergency Demand for Surgery," European Journal of Operational Research, vol. 185, pp. 1026-1037, 2008.

[4] M. Bruni, P. Beraldi, and D. Conforti, "A Stochastic Programming Approach for Operating Theatre Scheduling under Uncertainty," IMA Journal of Management Mathematics, vol. 26, pp. 99-119, 2015.

[5] M. Lamiri, F. Grimaud, and X. Xie, "Optimization Methods for a Stochastic Surgery Planning Problem," International Journal of Production Economics, vol. 120, pp. 400-410, 2009.

[6] M. Lamiri, J. Dreo, and X. Xie, "Operating Room Planning with Random Surgery Times," Automation Science and Engineering, 2007 CASE 2007 IEEE International Conference on, IEEE, 2007, pp. 521-526.

[7] D. Min and Y. Yih, "Scheduling Elective Surgery under Uncertainty and Downstream Capacity Constraints," European Journal of Operational Research, vol. 206, pp. 642-652, 2010.

[8] E. Hans, G. Wullink, M. van Houdenhoven, and G. Kazemier, "Robust Surgery Loading," European Journal of Operational Research, vol. 185, pp. 1038-1050, 2008.

[9] M. Heydari and A. Soudi, "Predictive/Reactive Planning and Scheduling of a Surgical Suite with Emergency Patient Arrival," Journal of medical systems, vol. 40, pp. 1-9, 2016.

[10] J. Beliën and E. Demeulemeester, "Building Cyclic Master Surgery Schedules with Leveled Resulting Bed Occupancy," European Journal of Operational Research, vol. 176, pp. 1185-1204, 2007.

[11] A. Fügener, E. W. Hans, R. Kolisch, N. Kortbeek, and P. T. Vanberkel, "Master Surgery Scheduling with Consideration of Multiple Downstream Units," European journal of operational research, vol. 239, pp. 227-236, 2014.

[12] S. Gul, B. T. Denton, and J. W. Fowler, "A Progressive Hedging Approach for Surgery Planning under Uncertainty," INFORMS Journal on Computing, vol. 27, pp. 755-772, 2015.

[13] J. J. Bisgaard, M. Heise, and C. Steffensen, "How Is Context and Context-Awareness Defined and Applied? A Survey of Context-Awareness," Department of Computer Science, Aalborg University, pp. 31-40, 2004.

[14] H. Lieberman and T. Selker, "Out of Context: Computer Systems That Adapt to, and Learn from, Context " IBM Systems Journal, vol. 39, pp. 617 - 632, 2000.

[15] B. Schilit, N. Adams, and R. Want, "Context-Aware Computing Applications," Mobile Computing Systems and Applications, 1994 Proceedings, Workshop on, 1994, pp. 85-90.

[16] A. K. Dey, "Understanding and Using Context," Personal and ubiquitous computing, vol. 5, pp. 4-7, 2001.

[17] P. Dourish, "What We Talk About When We Talk About Context," Personal and ubiquitous computing, vol. 8, pp. 19-30, 2004.

[18] G. D. Abowd, A. K. Dey, P. J. Brown, N. Davies, M. Smith, and P. Steggles, "Towards a Better Understanding of Context and Context-Awareness," Proceedings of the 1st international symposium on Handheld and Ubiquitous Computing, Springer-Verlag, 1999, pp. 304-307.

[19] J. E. Bardram, T. R. Hansen, and M. Soegaard, "Awaremedia: A Shared Interactive Display Supporting Social, Temporal, and Spatial Awareness in Surgery," Proceedings of the 2006 20th anniversary conference on Computer supported cooperative work, ACM, 2006, pp. 109-118.

[20] A. Hristova, "Conceptualization and Design of a Context-Aware Platform for User Centric Applications," Master of Science in Communication Technology Master, Department of Telematics, Norwegian University of Science and Technology, Trondheim, 2008.

Toward Robust Heart Failure Prediction Models Using Big Data Techniques

Heba F. Rammal
Information Technology Department
King Saud University
Riyadh, Saudi Arabia
Email: hrammal@ksu.edu.sa

Ahmed Z. Emam
Information Technology Department
King Saud University, Riyadh, Saudi Arabia
Computer Science and Math Department,
Menoufia University, Egypt
Email: aemam@ksu.edu.sa

Abstract— **Big Data technologies have a great potential in transforming healthcare, as they have revolutionized other industries. In addition to reducing the cost, they could save millions of lives and improve patient outcomes. Heart Failure (HF) is the leading death cause disease globally. The social and individual burden of this disease can be reduced by its early detection. However, the signs and symptoms of HF in the early stages are not clear, so it is relatively difficult to prevent or predict it. The main objective of this paper is to propose a model to predict patients with HF using a multi-structure dataset integrated from various resources. The underpinning of our proposed model relies on studying the current analytical techniques that support heart failure prediction, and then build a model based on Big Data technologies. To achieve this, we extracted different important factors of heart failure from King Saud Medical City (KSUMC) system, Saudi Arabia, which are available in structured, semi-structured and unstructured format. Unfortunately, a lot of information is buried in unstructured data format. We applied some pre-processing techniques to enhance the parameters and integrate different data sources in Hadoop Distributed File System (HDFS). Then, we applied data-mining algorithms to discover patterns in the dataset to predict heart risks and causes. Finally, the analyzed report is stored and distributed to get the insight needed from the prediction.**

Keywords- Big Data; Hadoop; Healthcare; Heart Failure; Prediction Model.

I. Introduction

In recent years, a new hype has been introduced into the information technology field called 'Big Data'. Big Data offers an effective opportunity to manage and process massive amounts of data. A report by the International Data Corporation (IDC) [1] found that the volume of data the whole humanity produced in 2010 was around 1.2 Zettabytes, which can be illustrated physically by having 629.14 Million 2 Terabytes external hard drives that can fill more than 292 great pyramids. It has been said that 'data is the new oil', so it needs to be refined like the oil before it generates value. Using Big Data analytics, organizations can extract information out of massive, complex, interconnected, and varied datasets (both structured and unstructured) leading to valuable insights. Analytics can be done on big data using a new class of technologies that includes Hadoop [2], R [3], and Weka [4]. These technologies form the core of an open source software framework that supports the processing of huge data sets. Like any other industry, healthcare has a huge demand to extract a value from data. A study by McKinsey [5] points out that U.S. spend at least 600$ - 850$ billion on healthcare. The report points to the healthcare sector as potential field where valuable insights are buried in structured, unstructured, or highly varied data sources that can now be leveraged through Big Data analytics. More specifically, the report predicts that if U.S. healthcare could use big data effectively, the hidden value from data in the sector could reach more than 300$ billion every year. Also, according to the 'Big Data cure' published last March by MeriTalk [6], 59% of federal executives working in healthcare agencies indicated that their core mission would depend on Big Data within 5 years.

One area we can leverage in healthcare using Big Data analytics is Heart Failure (HF); HF is the leading cause of death globally. It is the heart's inability to pump a sufficient amount of blood to meet the needs of the body tissues [7]. Despite major improvements in the treatment of most cardiac disorders, HF remains the number one cause of death in the world and the most critical challenges facing the healthcare system today [8]. A 2015 update from the American Heart Association (AHA) [9] estimated that 17.3 million people die due to HF per year, with a significant rise in the number to reach 23.6 million by 2030. They also reported that the annual healthcare spending would reach $320 billion, most of which is attributable to hospital care. According to World Health Organization (WHO) statistics [10], 42% of death in 2010 (42,000 deaths per 100,000) in the Kingdom of Saudi Arabia (KSA) were due to cardiovascular disease. Also, in KSA, cardiovascular diseases represent the third most common cause of hospital-based mortality, second to accidents and senility.

HF is a very heterogeneous and complex disease which is difficult to detect due to the variety of unusual signs and symptoms [11]. Some examples of HF risk factors are: breathing, dyspnea, fatigue, sleep difficulty, loss of appetite, coughing with phlegm or mucus foam, memory losses, hypertension, diabetes, hyperlipidemia, anemia, medication, smoking history and family history. Heart failure diagnosis is typically done based on doctor's intuition and experience rather than on rich data knowledge hidden in the database which may lead to late diagnosis of the disease. Thus, the effort to utilize clinical data of patients collected in databases to facilitate the early diagnosis of HF patients is considered a challenging and valuable contribution to the healthcare sector. Early prediction avoids unwanted biases, errors and excessive medical costs, which improve quality of life and services provided to patients. It can identify patients who are at risk ahead of time and therefore manage them with simple interventions before they become chronic patients. Clinical data are available in the form of complex reports, patient's medical history, and electronics test results [12]. These medical reports are in the form of structured, semi-structured and unstructured data. There is no problem to use structured data for risk prediction model. But, there is a lot of valuable information buried in unstructured data format because this data is very discrete, complex, multidimensional and noisy [13]. In our study, we collected patient's reports from a well-known hospital in Saudi Arabia: Kind Saud University Medical City (KSUMC).

The objective of our research is to mine the useful information from these reports with the help of cardiologists and radiologist to design a predictive model that will give us the prediction of HF. The paper is organized as follows. Section II introduces the related work. Section III describes the proposed architectural model and each process involved. In Section IV, the proposed research methodology is explained. The conclusion and future work of this research are found in Section V.

II. LITERATURE REVIEW

Big Data predictive analytics represents a new approach to healthcare, so it does not yet have a large or significant footprint locally or internationally. To the best of our knowledge, no prior work has investigated the benefits of Big Data analytics techniques in heart failure prediction problem. A work by Zolfaghar et al. [14] proposed a real-time Big Data solution to predict the 30- day Risk of Readmission (RoR) for Congestive Heart Failure (CHF) incidents. The solution they proposed included both extraction and predictive modeling. Starting with the data extraction, they aggregate all needed clinical & social factors from different resources and then integrated it back using a simple clustering technique based on some common features of the dataset. The predictive model for the RoR is formulated as a supervised learning problem, especially binary classification. They used the power of Mahout as machine learning based Big Data solution for the data analytics. To prove quality and scalability of the obtained solutions they conduct a comprehensive set of experiments and compare the resulted performance against baseline non-distributed, non-parallel, non-integrated dataset results previously published. RoR for CHF gained the interest of researchers due to their negative impacts on healthcare systems' budgets and patient loads. Thus, the development of predictive modeling solutions for risk prediction is extremely challenging. Prediction of RoR was addressed by, Vedomske et al. [15], Shahet al. [16], Royet al. [17], Koulaouzidis et al. [18], Tugermanet al. [19], and Kang et al. [20]. Although our studied problem is fundamentally different, as they are all using structure data; nevertheless, our proposed model could benefit from the proposed large-scale data analysis solutions.

Panahiazar et al. [21] applied decision trees, Random Forests, Adaboost, SVM and logistic regression to a dataset extracted from the EHR of the Mayo Clinic. The dataset included 5044 HF patients admitted to the Mayo Clinic from 1993 to 2013. For each patient, 43 predictor variables, expressing demographic data, vital measurements, lab results, medication, and co-morbidities, were recorded. The class variable corresponded to mortality status, consequently, three versions of the dataset were created, each one corresponding to survival period (1-year, 2-year, 5-year). 1560 instances out of 5044 were used for training and the rest 3484 instances for testing. The authors observed that logistic regression and Random Forests were more accurate models compared to others, also, among the scenarios, the best prediction accuracy was 87.12%.

Saqlain et al. [22] worked on 500 HF patients from the Armed Forces Institute of Cardiology (AFIC), Pakistan, in the form of medical reports. They started by manually applying pre-processing steps to transform unstructured reports into the structured format to extract data features. Then they perform multinomial Naïve Bayes (NB) classification algorithm to build 1-year or more survival prediction model for HF diagnosed patients. The proposed model achieved an accuracy and Area under the Curve (AUC) of 86.7% and 92.4%, respectively. Even though the above model is based on some attributes extracted from the unstructured data, they used a manual approach to achieve this. On the other hand, our model deals with unstructured data by automatically recognizing attributes using Machine Learning (ML) approaches without the need for a radiologist opinion. A scoring model for HF diagnosis based on SVM were proposed by Yang, G. et al. [23]. They applied it to a total of 289 samples clinical data collected from Zhejiang Hospital. The sample was classified into three groups: healthy group, HF-prone group, and HF group. They compared their results to previous studies which showed a considerable improvement for HF diagnosis with a total accuracy of 74.44%. Especially in HF-prone group, accuracy reaches 87.5%, and this implies that the proposed model is feasible for early diagnosis of HF. However, accuracy in the HF group is not satisfactory due to the absence of symptoms and signs and also due to the high prevalence of conditions that may mimic the symptoms and signs of heart failure.

More studies are listed in Table 1, which were collected and summarized as recent analytics techniques and platform to predict heart failure. The table shows that supervised learning technique is the most domainant techniques in building HF predication model, also Weka and Matlab are the preferable platforms to build HF prediction model.

The literature presented above shows a gap in multi-structured predictors for HF prediction and data fusion which will be our main task. It is easy to observe that our effort is orthogonal to this related work but, unlike us, none of these works deal with the problem semi-structured or unstructured HF predictor variable. They did not generate Big Data analytics predication model, nor do they perform on large scale or distributed data.

TABLE 1. STATE OF ART FOR HF PREDICTION STUDIES

Author	Prediction Technique Used		Platform	Objective
Zolfaghar et al. (2013)	**Supervised learning**	Logistic regression, Random forest	Mahout	BD solution to predict the 30- day RoR of HF
Meadam et al. (2013)		Logistic regression, Naive Bayes, Support Vector Machines	R	Evaluation preprocessing techniques for Prediction of RoR for CHF Patients
Yang et al. (2010)		support vector machine (SVM)	n/a	A heart failure diagnosis model based on support vector machine
Panahiazar et al. (2015)		Decision trees, Random Forests, Adaboost, SVM and logistic regression	n/a	Using EHRs and Machine Learning for Heart Failure Survival Analysis
Donzé et al. (2013)		Cox proportional hazards	SAS	Avoidable 30-Day RoR of HF
Zolfaghar et al. (2013)		Naive Bayes classifiers	R	Intelligent clinical RoR of HF calculator
Bian et al. (2015)		Binary logistic regression	n/a	Scoring system for the prevention of acute HF
Suzuki et al. (2012)		logistic regression	SPSS	Scoring system for evaluating the risk of HF
Auble et al. (2005)		Decision tree	SPSS	Predict low-risk patients with HF
Pocock et al. (2005)		Cox proportional hazards	n/a	Predictors of Mortality and Morbidity in patients with CHF
Miao, Fen et al. (2014)		Cox proportional hazards	R	Prediction for HF incidence within 1-year
Dangare et al. (2012)		Decision Trees, Naïve Bayes, and Neural Networks	Weka	HD prediction system using DM classification techniques
Rupali R. Patil (2014)		Naive Bayes classifiers	MATLAB	HD prediction system
Rupali R. Patil (2012)		Artificial Neural network	Weka	A DM approach for predication of HD
Wu, Jionglin et al. (2010)		Logistic regression, SVM, and Boosting	SAS, R	HF prediction modeling using EHR
Zebardast et al. (2013)		Generalized Regression Neural Networks	MATLAB	Diagnosing HD
Vanisree K. & Singaraju J. (2011)		Multi layered Neural Network	MATLAB	Decision Support System for CHD Diagnosis
Guru et al. (2007)		Neural network	MATLAB	HD prediction system
R, Chitra and V, Seenivasagam (2013)		Cascaded Neural Network	n/a	HD Prediction System
Sellappan Palaniappan and Rafiah Awang (2008)		Decision trees, naïve bayed and neural network	.Net	HD prediction system using DM techniques
K. Srinivas et al. (2010)		Naive Bayes classifiers	Weka	DM technique for prediction of Heart Attacks
Saqlain et al. (2016)		Naive Bayes	n/a	Identification of HF using unstructured data of Cardiac Patients
Strove, Sigurd et al. (2004)	**Structured predication (Bayesian network)**		HUGIN	Decision Support Tools in Systolic HF Management
Gladence, L.M. et al. (2014)			Weka	Method for detecting CHF
Liu, Rui et al (2014)			MicrosoftAzure (R & python)	Framework to recommend interventions for 30-Day RoR of HF
C. Ordonez (2006)	**Association rules**		n/a	HD Prediction
M. Akhil Jabbar et al. (2012)	**Associative classification**	Gini index, Z-statics & genetics algorithm	n/a	Decision Support System for HD prediction
K. Chandra Shekar et al. (2012)		association rule mining and classification	Java	Algorithm for prediction of HD

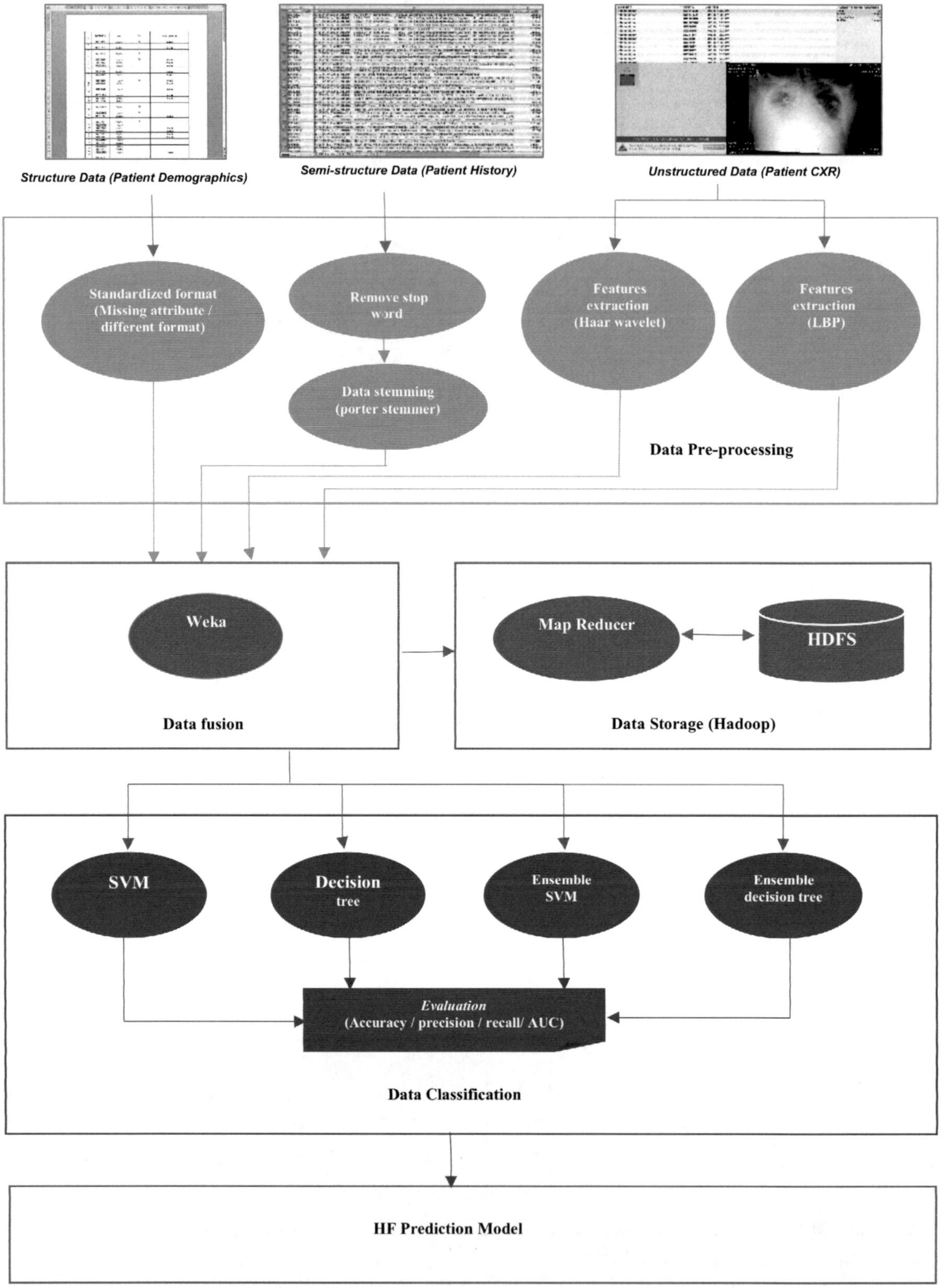

Figure 1. HF Prediction Model.

III. PROPOSED ARCHITECTURE

Predictive analysis can help healthcare providers accurately expect and respond to the patient needs. It provides the ability to make financial and clinical decisions based on predictions made by the system. Building the predictive analysis model includes various phases as mentioned in the literature. (Figure 1 shows the complete architecture of the proposed model).

IV. PROPOSED METHODOLOGY

In the following, we will describe the adaptive methodology and each step towards our proposed model.

A. *Data Collection*

In collaboration with King Saud Medical City (KSUMC) system located in Riyadh, Saudi Arabia all needed clinical and demographic data were adopted to evaluate the performance of the proposed hybrid method in identifying HF risk in patients. The dataset contained 100 real patient records extracted form KSUMC electronic health recod (EHR) with approval from KSUMC administrative office. The selected sample covers most distribution domain from different ages and different sociodemographic. At the same time, validation of the selected dataset was achieved by consolidating some cardiologist and data scientist. The selected dataset has many noises such as missing values and misidentified attributes. The output values were categorized into two labels denoted as Non-HF (meaning HF is absent) and HF (meaning HF is present). One of the major steps is the distillation of data, which responsible of determining the subset of attributes (i.e., predictor variables) that has a significant impact in predicting patient with HF from the myriad of attributes present in the dataset. In our study, the definitions and categories of the HF attributes are summarized in Table 2.

B. *Data Preprocessing*

In this phase structured, semi-structured, and unstructured data are accumulated, cleansed, prepared, and made ready for further processing.

TABLE 2. SELECTED ATTRIBUTED FROM THE DATASET

Label		Feature	Format
Structured	Demographics	Age	Numeral
		Sex	Binary
		Place of birth	Nominal
Semi-Structured	Clinical indications / History	Hypertension, Anemia, Diabetes, Chronic Kidney Disease, Ischemic heart disease, SOB, Swilling hands, Cough, Previous CHF	String
Un-Structured	Front CXR	64 Features (Haar)	Numeral
	Back CXR	61 Features (LBP)	
	Side CXR		

- Raw structured information has some missing values and written in different formats during information entry or management, we had to do screening before data analyzing. Those data with too many missing attributes were all wiped off when we selected the sample set. Also, all data formats were standardized, see Table 3.
- Apply text analysis techniques on the semi-structured dataset to get the needed information. Two steps were applied to the text to process the data, stop word removal and stemming. Stop word removal - illustrated in Figure 2 - helped in removing all common words, such as 'a' and 'the' from the text. Next, Porter algorithm was used as the stemmer to identify and remove the commoner morphological and inflexional endings from words [24].
- Extracting all needed features from the unstructured dataset, which includes 3 types of Chest X-Ray (CXR) images (front CXR, back CXR, and side CXR) using MATLAB. Haar wavelet and local binary pattern (LBP) were applied to over 150 CXR images. Haar was used since it is the fastest technique that can be used to calculate the feature vector [25]. This was performed based on applying the Haar wavelet four times to divide the input image into 16 sub-images, illustrated in Figure 3. 64 features that include Energy, Entropy, Homogenous as 3D XYZ were found using Wavelets features, see Figure 4.

  On the other hand, LBP has been found to be a powerful and simple feature yet very efficient texture operator which labels the pixels of an image by calculating each pixels' neighborhoods' thresholding then considers the result as a binary number. 61 features were found using LBP. Principle component analysis (PCA) was applied to properly rank and compute the weights of the features to find the most promising attributes to predicate HF from 64 / 61 features found. The selected attributes were used to train the classifiers to get a better accuracy.

TABLE 3. UNSTANDARIZED STRUCTRED DATA

	Age	Sex	P_B	Diagnosis
1	045Y	Female	Riyadh	HF
2	62	F	?	HF
.....	....	......	.....	.....
100	098	male	riy	Non-HF
	Explanatory Data			Label

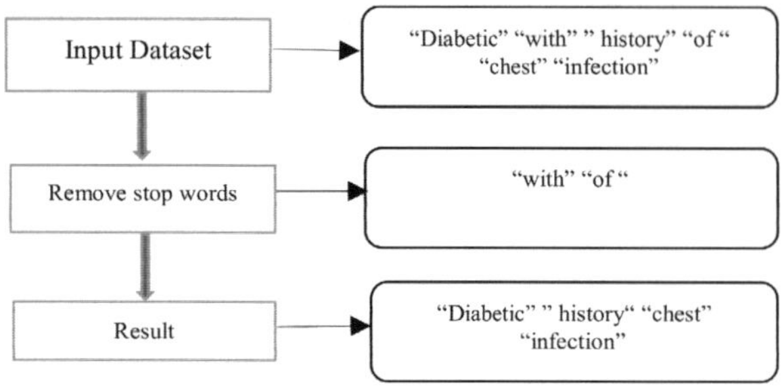

Figure 2. Stop word removal.

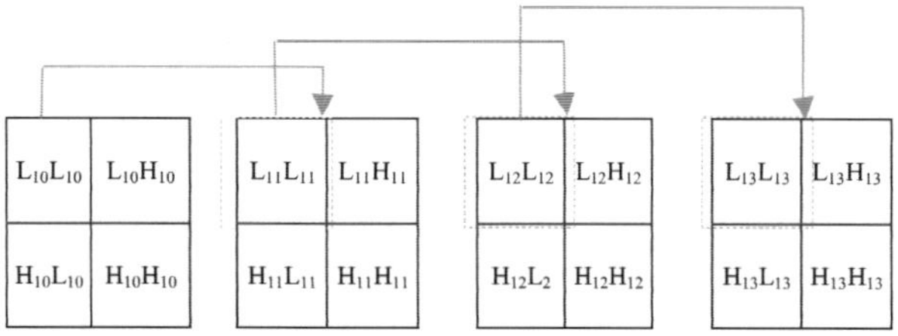

Figure 3. Applying haar wavelet four times.

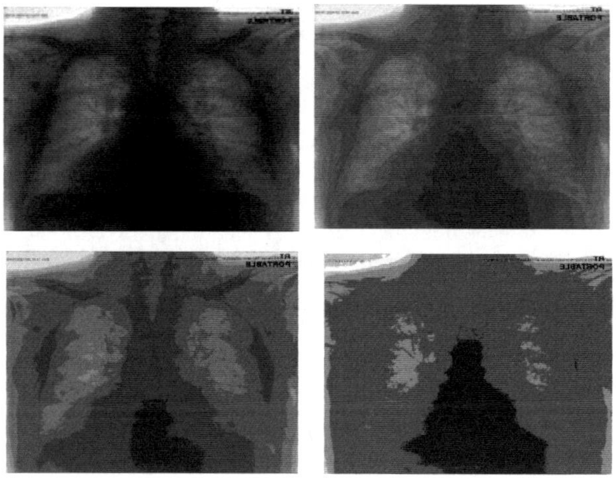

Figure 4. Result from wavelet.

C. *Data Storage and Fusion*

After pre-processing the data and extracting all the needed attributes, the statistics feature from CXR scan images with other attributes will be integrated using Big Data tools to generate the needed data that will be used for training and testing & finally produce the predictive model. We leverage the power of Hadoop as a framework for distributed data processing and storage. Hadoop is not a database, so it lacks functionality that is necessary for many analytics scenarios. Fortunately, there are many options available for extending Hadoop to support complex analytics, including real-time predictive models such as Weka (Waikato Environment for Knowledge Analysis), which we used in our study. We added distributedWekaHadoop to Weka, which works as a Hadoop wrapper for Weka.

D. *Data Classification*

In this study, each set of the data (Structured, Semi-structured, and Unstructured) trained and tested using data mining algorithms in Weka. Knowledgeflow was used in Weka which presents, a workflow inspired interface, see Figure 5. Data was trained using two state-of-the-art classification algorithms including, SVM and Decision Tree. In the end, accuracy, precision, recall and, Area under the Curve (AUC) were used as performance measures. We get all these measures by using confusion matrix because it contains all True Positive (TP), True Negative (TN), False Positive (FP) and False Negative (FN) assessments.

V. Conclusion And Future Work

Big Data analytics provides a systematic way for achieving better outcomes like availability and affordability of healthcare service to all members of the population. Non-Communicable Diseases like Heart Failure is one of the major health hazard in the KSA. By transforming various health records of HF patients into useful analyzed result, this analysis will make the patient understand the complications that can occur.

The literature shows a gap in multi-structured predictors for HF prediction and data fusion, which is our main task. It is easy to observe that our effort is orthogonal to this related work but, unlike us, none of these works deal with the problem semi-structured or unstructured HF predictor variable. They did not generate Big Data analytics predication model, nor do they perform on large scale or distributed data.

Combining structured dataset (sociodemographic), semi-structured dataset (cardiology diagnoses), and unstructured dataset (Patient X-ray images) is a very hard task. In this research, data fusion played a vital role in combining multi-structured datasets. The goal of this research deals with the study of HF predication in healthcare industry using Big Data analytics technique. The design of predictive analysis model of HF may give enhanced data and analytics for better results in healthcare.

As future work, we will use a larger dataset for training. We will also incorporate more medical data into the model, better simulating how a cardiologist makes a decision. Finally, we will use different data mining techniques to extract the buried information from the patient semi-unstructured /unstructured reports.

References

[1] J. Gantz and D. Reinsel, "The digital universe decade – are you ready?" External publication of IDC (Analyse the Future) information and data, pp. 1- 16, 2010.

[2] Hadoop Apache. Available at http://hadoop.apache.org/, Last accessed March 2017.

[3] R Project. Available at https://www.r-project.org/about.html, Last accessed March 2017.

[4] P. Navas, Y. Parra, and J. Molano, "Big Data Tools: Haddop, MongoDB and Weka". International Conference on Data Mining and Big Data, pp 449-456, 2017.

[5] McKinsey and Company, McKinsey Global Institute, Big Data: The next frontier for innovation, competition and productivity. Available at http://lateralpraxis.com/download/The_big_data_revolution_in_healthcare.pdf, Last accessed March 2017.

[6] Meritalk, The Big Data cure. Available at http://www.meritalk.com/bigdatacure, Last accessed March 2017.

[7] National Heart, Lung, and Blood Institute, What is heart failure. Available at http://www.nhlbi.nih.gov/health/health-topics/topics/hf, Last accessed April 2017.

[8] World Health Organization (WHO) (2015). Cardiovascular diseases (CVDs). Available at http://www.who.int/mediacentre/factsheets/fs317/en/ , Last accessed March 2017.

[9] American Heart Association. Heart Disease and Stroke Statistics – At-a-Glance. Available at http://www.heart.org/idc/groups/ahamah-public/@wcm/@sop/@smd/documents/downloadable/ucm_470704.pdf, Last accessed March 2017.

[10] Mistry Of Health (MOH), "Cardiovascular Diseases Cause 42% of Non-Communicable Diseases Deaths in the Kingdom". Available at https://www.moh.gov.sa/en/Ministry/MediaCenter/News/Pages/News-2013-10-30-002.aspx, Last accessed March 2018.

[11] Ishwarappa and J. Anuradha, "A Brief Introduction On Big Data 5Vs Characteristics And Hadoop Technology". Procedia Computer Science 48, pp. 319-324, 2015.

[12] J. A. Hartigan and M. A. Wong, "Algorithm AS 136: A k-means clustering algorithm,", Series C (Applied Statistics), vol. 28, pp. 100-108, 1979.

[13] E. AbuKhousa and P. Campbell, "Predictive data mining to support clinical decisions: An overview of heart disease prediction systems," Proc. IEEE, Innovations Information Technology (IIT), pp. 267-272, March 2012.

[14] K. Zolfaghar et al., "Big Data Solutions for Predicting Risk-of-Readmission for Congestive Heart Failure Patients", IEEE Inter. Conf. on Big Data, 2013.

[15] M. A. Vedomske, D. E. Brown and J. H. Harrison, "Random forests on ubiquitous data for heart failure 30-day readmissions prediction", Proceedings of the 12th international conference on machine learning and applications, vol. 2, pp. 415-421, 2013.

[16] S. J. Shah et al. "Phenomapping for novel classification of heart failure with preserved ejection fraction". Circulation. Vol 131, pp. 269–279, 2015.

[17] S. B. Roy, A. Teredesai, Zolfaghar K., Liu R., and Hazel D., "Dynamic hierarchical classification for patient risk-of-readmission". Proceedings of the 21th ACM SIGKDD international conference on knowledge discovery and data mining, pp. 1691–1700, 2015.

[18] G. Koulaouzidis, D. K. Iakovidis, and A. L. Clark, "Telemonitoring predicts in advance heart failure admissions". Int J Cardiol, vol. 216, pp. 78–84, 2016.

[19] L. Turgeman and J. H. May, "A mixed-ensemble model for hospital readmission.", Artif Intell Med, vol. 72, pp. 72–82, 2016.

[20] Y. Kang, M.D. McHugh, J. Chittams and K.H. Bowles, "Utilizing home healthcare electronic health records for telehomecare patients with heart failure. A decision tree approach to detect associations with rehospitalizations". Comput Inform Nurs, vol. 34 no. 4, pp.175–182, 2016.

[21] M. Panahiazar, V. Taslimitehrani, N. Pereira and J. Pathak, "Using EHRs and machine learning for heart failure survival analysis." Stud Health Technol Inform, vol. 216, pp. 40–44, 2015.

[22] M. Saqlain, W. Hussain, N. Saqib and A. Muazzam Khan, "Identification of Heart Failure by Using Unstructured Data of Cardiac Patients.", 45th International Conference on Parallel Processing Workshops, 2016.

[23] G. Yang et al., "A heart failure diagnosis model based on support vector machine". 3rd International Conference on Biomedical Engineering and Informatics. 2015;

[24] C. Moral, A. Antonio, R. Imbert and J. Ramírez, "A survey of stemming algorithms in information retrieval." Information research, vol. 19 no. 1, March 2014.

[25] S. Arora, Y. Brar and S. Kumar, "HAAR wavelet transform for solution of image retrieval." International Journal of Advanced Computer and Mathematical Sciences ISSN, vol. 5, no. 2, pp 27-3, 2014

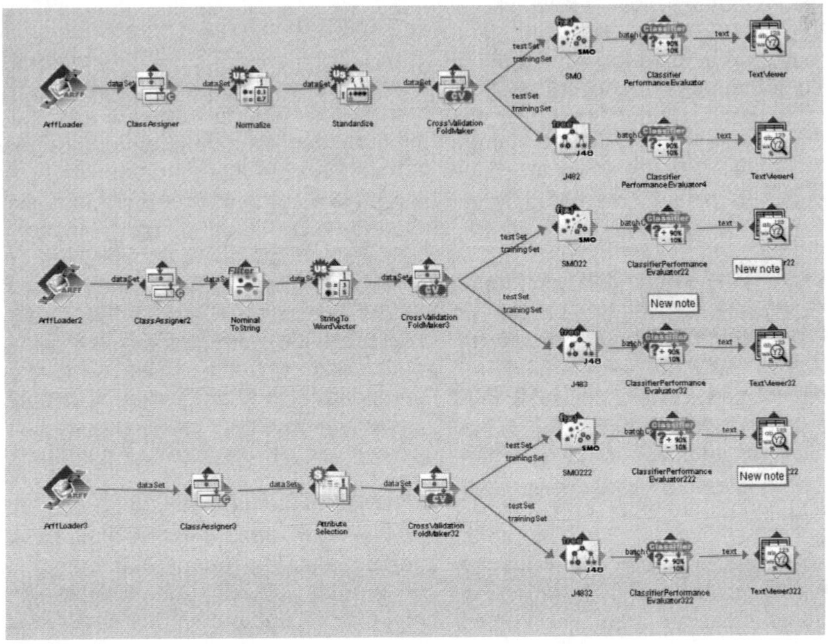

Figure 5. The Proposed Knowledgeflow using Weka.

An Expert System Framework for Lifestyle Counselling

Benedek Szakonyi, Tamás Lőrincz, Ágnes Lipovits, István Vassányi
Faculty of Information Technology, University of Pannonia,
Veszprém, Hungary
email: benedek.szakonyi@virt.uni-pannon.hu, tamas.lorincz@virt.uni-pannon.hu,
lipovitsa@almos.uni-pannon.hu, vassanyi@almos.vein.hu

Abstract— **The implementation and support of proper lifestyle is a key factor in preventing chronic diseases and also in the management of already manifested ones. This paper presents a CLIPS (C Language Integrated Production System) based framework that can be used to develop expert systems for assisting users to learn a proper lifestyle. The relevance and applicability of this system is shown in the domains of diet and exercise based lifestyle management and workplace related ergonomics. Preliminary results show that the framework can properly evaluate the formulated basic knowledge bases and is suitable for the tasks intended.**

Keywords – CLIPS; expert system; lifestyle assessment; telemedicine; work ergonomics.

I. INTRODUCTION

Lifestyle is a key factor in the prevention and management of chronic diseases [1]. A shortage of manpower in professions aiding people to "live a healthy life" (e.g., doctors, dietitians, ergonomists, etc.) have surfaced recently, and is growing gradually, as the number of treatments to be done is increasing faster than that of such experts [2]. A Lifestyle Coach Framework, aimed at improving people's way of living, was developed for creating expert systems that could be used in solving these problems. With this framework, targeting general audiences is possible, just as well as focusing on specific patient groups, like diabetics. Two areas, logging-based diet and exercise oriented assessment, and workplace ergonomics were investigated. Their common ground is that for both, evaluation of user behavior can be achieved by analyzing events with expert systems (which differ only in the knowledge used).

Expert systems (ES) have been used in many application areas (e.g., operations, production, finance, etc.), healthcare and medical cases included. From the different approaches on the demanding task of how to transform human knowledge into computer comprehensible data, the most dominant solution has been applying a rule-based system [3]. Interestingly, the aforementioned research inspected about 300 expert systems from the last 30 years, yet it seems, that none of them addressed the problem of daily lifestyle assessment.

In the treatment of diseases affecting considerable masses in modern societies, changing lifestyle was identified as a key component. Carelessness has a huge impact on life quality, as developing harmful conditions is strongly related to living unruly, unhealthy lifestyles. The Lavinia Lifestyle Application [4] serves as a tool for users to log their daily nutrition and medicine intake, and their physical activity. Analyzing the data acquired in this way, by using the knowledge of dietary experts, could improve the effective assessment of users' lifestyles.

Computer based work, already in the early years of the 2000s, was found to be dominant in developed countries [5]. This extensive use of computers results in workers sitting for most of their shifts [6], but research has shown, that by investigating working environments and improving workplace ergonomics, the negative effects of this prolonged sedentariness can be reduced effectively [7]. Yet, in that case, participants were in direct contact with (ergonomist) experts, which, although desirable, on global scale could be found infeasible and/or unprofitable, so computerizing and automating this procedure could serve as a good alternative.

Section II details the methods used, Section III describes the framework developed, and in Section IV the two modules are presented. In Section V, preliminary results are shown, while Section VI addresses the conclusions and future work.

II. METHODS

A. *Expert Systems*

There are different approaches on implementing expert systems (case-based reasoning, fuzzy logic, genetic algorithms, etc.), but for lifestyle assessment a rule-based ES is sufficient, as the experts of the related fields (e.g., dietitians, ergonomists) can usually express their knowledge in the form of IF-THEN like rules. Rules are applied on facts (statements describing the current state of the system, and the new events) by an inference engine, to deduce new facts and provide explanation, if needed. An advantage of using such an expert system is that updating the knowledge base can be easily done, as it is separated from the reasoning engine, so changes in it will not affect basic functionality of the system.

The sources of the rules used in the lifestyle management module are dietitian experts who participated in the clinical trials related to the development of Lavinia, as there are numerous different recommendations and standards to otherwise choose from. For evaluating user postures, RULA

(Rapid Upper Limb Assessment) [8], a generally accepted scoring method for measuring physical load on office workers, is selected. One of its advantages is that it rates not just the postures themselves, but also considers how much time is spent in them.

B. *Modelling Time*

For many problems addressed by expert systems, the temporal relations of facts are not relevant. In lifestyle assessment, however, time has an impact on making decisions: for example, a fact like "the user has exercised today" should be handled differently, whether if it is the first or the 7th occasion in a week (as the latter is a greater achievement). Handling time passing is crucial, as there may be facts that have lost their actuality and thus should be updated or removed, while others may have to be inserted after specific time intervals have passed: a fact like "the user ate fish today" should change to "the user ate fish yesterday" on the next day.

This "time problem" has another aspect: the time concept for logging events is different from the one used by experts. The former is assigning a specific timestamp to every event, while experts tend to refer to events from the present time (e.g., "Yesterday the patient ran 5 km"). To solve this semantic gap, the framework uses relative timestamps (fact ages). When inserted, facts have an age of 0, which then can be updated by aging rules as many times as needed. These rules activate (fire) when given signaling facts (that indicate the passing of a specific time interval) are inserted. In order to allow the usage of various measures of time (e.g., week, day, hour), the framework is currently able to handle two different time dimension configurations (one for each module created). The first one is for applications in which the shortest interval required is one hour, and longest is a week. This case, used in Section IV A, has time units "hour", "day" and "week". The other one is suitable when the minimal interval is one minute, and the maximal does not exceed a day. This setting, applied in Section IV B, offers two units: "minute" and "hour".

C. *Validation and Testing*

Once a sufficient number of rules have been formulated, their proper validation is needed, by field tests, if possible, or by using Turing test-like blind evaluation, when both the system's and expert's outputs for the same input are evaluated by other experts, without knowing who is performing.

For expert systems, simulation of user behavior is needed to provide the necessary input. Using time configurations allows accelerating this evaluation process: after inserting the relevant facts for a given time interval (minute, day, etc.), a fact stating that the current interval is over can be inserted, and the test can continue with the next one. This way, even data spanning a year can be processed in some minutes.

For the lifestyle management module, former user logs of the Lavinia Lifestyle Application are available and can be used for testing. For the workplace ergonomics part, video recordings from office work surveillance/monitoring could be used for the same purpose.

III. THE FRAMEWORK

From the available expert system solutions, CLIPS (C Language Integrated Production System) [9] was chosen to serve as a basis, as it has already proved its value (e.g., in NASA's space shuttle missions [10], security purposes [11]), and for its advantages such as being a public domain software, having low resource needs and for supporting portability (C code). With regard to the portability offered, the framework is based on C++ language. An overview of the framework developed is shown in Figure 1.

The CLIPS runtime is responsible for evaluating rules and facts. The event dispatcher inserts the incoming facts from the interface into CLIPS and to the database (DB). It also stores the output of CLIPS (facts deducted) in the database, and based on it, forwards any messages to the interface, if needed. The database is not used for only modelling the problem, it serves as a backup on CLIPS' work memory. It can be used for initialization, when a former state of the system is to be reconstructed, and for debugging purposes, to investigate the validity of the reasoning. The main entities of this database are the following:

- *rule*, *fact* and *event* tables – containing the defined rules, facts and events
- *rule_fact* table – storing the connection of facts and the rules that deduced them.
- *rset* and *rule_set* tables – allowing rules to be sorted into groups. This provides the ability to change which rules are currently used. E.g., apply different rules for diabetics and healthy people.
- *rule_gui* table – storing the messages to be displayed, for given rules.

The interface (using Google Protobuf [12]) is for other applications to use, when modifications in the expert system are needed, and to receive its responses. There are two groups of Protobuf messages: *control* and *event*. *Control* messages can be used to manage the expert system, e.g., *INTERVAL_PASSED* "orders" the expert system to age facts by the shortest time unit configured. An *event* can be of two types, *fact* or *consequence*: the former is for inserting new facts (module specific system input), the latter is for outgoing communication (e.g., displaying messages). The structure of Protobuf messages is shown in Figure 2, where the possible fact values are for the lifestyle management module.

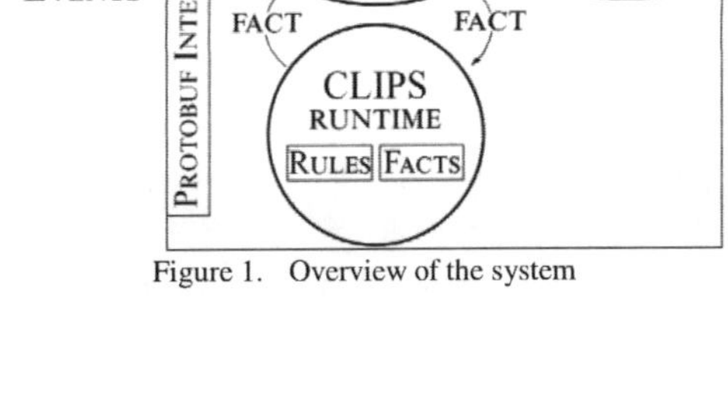

Figure 1. Overview of the system

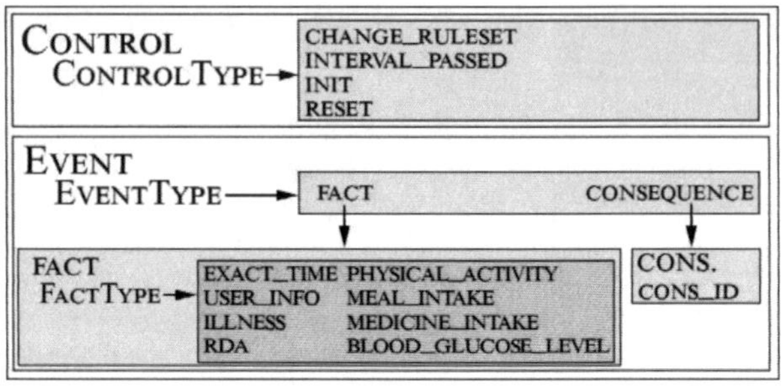

Figure 2. Interface message structure (lifestyle management module)

For the work ergonomics module, a similar structure is used, with the same *controls* and *event* types, but of course, it uses different *fact* types that are relevant to the ergonomic evaluation of user behavior.

IV. APPLICATION DOMAINS

A. *Lifestyle Management*

To analyze the user logs provided by Lavinia, an expert system module was developed with the framework presented in Section III, that applies rules formulated by dietitians.

An example for such a rule is "For exercising on every day in the last week, the user is to be complimented". Defining a rule checking the last 7 days for exercises done is simple with this framework: when the sufficient number of facts having the required ages (0, 1, …, 6 days) is found, the compliment is sent. (This, of course, requires logging the exercises and forwarding this information to the system.) To avoid bombing the user with the same message repeatedly (as the notification requirements are met), after the rule has fired and the first compliment is sent, a new fact stating the completion of the feedback is deducted (see Figure 3). This rule, however, is to be removed after a specific time interval, to make complimenting possible again. This, as the new rule by definition has an age when created, is achieved by adding a simple removal rule.

Another rule, requiring a different approach, is "If user's daily fiber intake reaches the recommended value, a compliment is to be sent". As users' goals can differ concerning their weight plan (gaining, losing or maintaining), and with them their corresponding recommended fiber intakes, rules handling all possibilities were created. After each midnight, the recorded intake value is set to zero, and when meals are logged, it is increased accordingly to the food's fiber amount.

```
(defrule physical_activity_whole_week
   (physical_activity_whole_week)
   (not (physical_activity_feedback
          (fact_age_day ?g&:(= ?g 0))))
   =>
   (clips_event_happened 5)
   (assert (physical_activity_feedback))
)
```

Figure 3. A rule for sending a motivational message

Notifying users when the daily intake is reached is similar in each cases, but the related reactions are distinct: it is either a warning to avoiding additional consumption (maintaining or losing weight) or a compliment (weight gain).

To validate the rules created, former logs of randomly selected Lavinia users were used to simulate real human behavior (see Section V).

B. *Work Ergonomics*

The problem of assessing office work ergonomics can be separated into two subtasks: the first is the already discussed problem of translating human knowledge for computers. The second is providing the means for gathering the user input to be analyzed, without requiring active cooperation from users.

In case of ergonomic assessment, rules focus mostly on the users' body postures, usually with regard to the time spent in those poses. This means concerning postures both for the whole body (e.g., "the users should not sit in one place for too long"), and in more detail, like "for most of the time, users' forearms should be properly supported", or "the users should not spend too much time with their heads tilted (cradling a phone)". Of course, as previously mentioned, time concepts such as "too much", "too long", etc., are incomprehensible for computers, so ergonomist experts are needed to define more precise time intervals. Moreover, they are the ones who can specify what is an acceptable posture and what is not for rules such as "the users should sit with their back held properly". However, this precision would prove useless, if the other subtask, collecting posture data, is not solved correctly.

A trivial approach for monitoring users' postures is using some camera equipment. To precisely detect user position or posture, however, a "traditional" 2 dimensional (2D) device is insufficient, as in this case, 3 dimensional data is required. Of course, using more than one 2D camera could solve this problem, but applying a single sensor with the required functionality is preferable. Such a device is the Microsoft Kinect v2, that provides video (colored and infrared) and depth data streams, as well as a built-in feature for body recognition and tracking (25 skeletal joins per person, up to 6 people) [13][14].

The accuracy of Kinect in office monitoring was addressed in [15], showing that its tracking capabilities are acceptable, but whether the received data is usable for ergonomic evaluation or not, was not mentioned. To investigate this, a test software, measuring and displaying the acquirable information of user behavior, has been created, and is currently being evaluated by ergonomists.

Parallel to this, with the framework developed, a basic expert system module was created, in order to validate if it can react properly to simple workplace related situations or not. For this reason, some basic ergonomic rules were formulated, these are detailed in Section V.

V. RESULTS

In total, 29 rules were formulated so far in the lifestyle managing module: 6 of them are automatically generated by the framework for aging facts, the rest serve for lifestyle assessment. The implemented statements given by dietitians were:

- When users log their fasting blood glucose level, they are to be rewarded.
- When users log their insulin dosage, they are to be rewarded.
- Walking 30 minutes a day should be rewarded.
- Users should be complimented if, for the last week, they exercised every day.
- If the user reached the recommended daily fiber intake, a compliment is to be sent.
- The users are to be complimented, if their BMI (Body Mass Index) drops below 25.
- Logging meals containing fish dishes on two different days in the last week is to be complimented.

The tests have shown that the already defined rules provide the required functionality and work properly, so work on implementing additional rules to cover a wider part of lifestyle assessment can begin. Moreover, trials of running the framework on smartphones were successful, and have shown that it is possible to use this lifestyle assessing system on mobile devices.

The work ergonomics module was tested in a similar fashion, but as evaluation of the data acquiring sensor is still in progress, plain concepts and lenient constraints were used in rules like

- After 90 minutes of continuous sitting, the user is to be advised to stand up/take a walk.
- After 3 hours of continuous sitting, the user is to be compelled to stand up/take a walk.
- If for at least half of the time in the last 60 minutes, the angle of the user's back and the vertical position was more than 30 degrees, a warning is to be sent.
- If for the last 60 minutes, it occurred more than 3 times that the user spent more than 3 minutes with their head rotated more than 30 degrees (compared to the vertical position), a warning is to be sent-

As in the case of the other module, tests have proved that the basic functionality of this module is satisfactory, and can be used for implementing concepts regarding work ergonomics, however, it is clear that more precise intervals should be used in discriminating proper and improper postures.

Initial results on Kinect's applicability seem to confirm that the device is suitable for ergonomic evaluation, but only after some preprocessing done, as in some occasions, false body detections and strange body pose predictions have been observed. Changing the position and orientation of the device was found to have a notable effect on how accurate the readings are, which correlates to the findings of Wiedemann et al. [15].

VI. CONCLUSION AND FUTURE WORK

This paper presented a framework for creating expert systems for lifestyle assessment, and two examples for its application domains were introduced.

The tests show the developed framework and the modules to be promising, however, further study has to be done on how user behavior can be assessed more thoroughly, and on what the related system reactions should be (what other rules to implement). Based on these findings, if needed, the framework itself should be improved accordingly.

For the lifestyle management module, apart from additional rules, further improving the method for providing positive feedback on Lavinia users' behavior (what the messages should be) is also important, as this can have a great impact on their motivation and success.

The implementation of the RULA-based rule set is in progress, and a total of 30-40 rules is expected to cover all significant aspects of work ergonomics. The validation of the developed module, as well as its rule-base, is to be completed in early 2018.

In both testing procedures of the work ergonomics module, the idea of having the system running on a different computer than of the user's has emerged. This would save the user's device from providing the computational requirements of the data processing, could strengthen error resistance (e.g., protection from power outage related damage) and would allow additional security measures (limit access to the system).

However, an additional personal computer in the work environment might be found unappealing (e.g., it requires more space, cabling, etc.). A solution to this problem, and also to the proper placing of the monitoring sensor, could be using an "intelligent workstation": a mechanical desk, embedded with a single-board computer capable of executing the tasks related to data acquisition and expert system management, with a structure that makes it possible to install the sensor in an optimal position. Moreover, other useful features could also be implemented, e.g., motorized height adjustment of the desk or a self-adjusting adaptive desk lamp. On how these could be efficiently implemented, and what other handy utilities could be used, further research is needed.

ACKNOWLEDGMENT

This research was supported by the ÚNKP-17-2 New National Excellence Program and the NTP-HHTDK-17-0033 Program of the Ministry of Human Capacities, Hungary, and was partially supported by the NKFI-6 fund through project K120369 (OTKA).

REFERENCES

[1] F. Sassi and J. Hurst, "The prevention of lifestyle related chronic disease: an economic framework", OECD Health Working Papers, p. 32, 2008.

[2] World Health Organization and Global Health Workforce Alliance, "A universal truth: no health without a workforce", Third Global Forum on Human Resources for Health Report, November 2013, ISBN 978 92 4 150676 2.

[3] W. P. Wagner, "Trends in expert system development: A longitudinal content analysis of over thirty years of expert system case studies", Expert Systems with Applications, vol. 76, Pages 85-96, 2017, doi: 10.1016/j.esq.2017.01.028.

[4] I. Kósa et al., "A fast, Android based dietary logging application to support the life style change of cardio-metabolic patients", Global Telemedicine and eHealth Updates: Knowledge Resources, vol. 7, pp. 553-556, 2014, ISSN 1998-5509.

[5] M. J. Smith, F. T. Conway, and B. T. Karsh, "Occupational stress in human computer interaction". Industrial Health, vol. 37, pp. 157–173, 1999.

[6] J. O. Hill and J. C. Peters, "Environmental contributions to the obesity epidemic", Science, vol. 280, pp 1371–1374, 1998.

[7] M. M. Robertson, Y. H. Huang, and J. Lee, "Improvements in musculoskeletal health and computing behaviors: Effects of a macroergonomics office workplace and training intervention", Applied Ergonomics, vol. 62, pp. 182-196, Jul. 2017, doi:10.1016/j.apergo.2017.02.017.

[8] R. Lueder and N. Corlett, "A proposed RULA for computer users", Proc. Ergonomics Summer Workshop, UC Berkeley, Aug. 1996, pp. 8-9.

[9] Gary Riley, CLIPS - A Tool for Building Expert Systems, www.clipsrules.net/?q=AboutCLIPS [retrieved: Feb, 2018].

[10] J. F. Muratore, T. A. Heindel, T. B. Murphy, A. N. Rasmussen, and R. Z. McFarland, "Applications of artificial intelligence to space shuttle mission control", Proc. Innovative Applications of Artificial Intelligence, 1989, pp. 105-110, ISBN 978-0-262-69137-6.

[11] D. Silberberg and R. Thomas, "The NASA personnel security processing expert system", Proc. Innovative Applications of Artificial Intelligence, 1996, pp. 1527-1535, ISBN 978-0-262-51091-2.

[12] Google Inc., Protocol Buffers, developers.google.com/ protocol-buffers [retrieved: Feb, 2018].

[13] Z. Zhang, "Microsoft Kinect sensor and its effect", IEEE MultiMedia, vol. 19, no. 2, pp. 4-10, Feb. 2012., doi: 10.1109/MMUL.2012.24.

[14] E. Cippitelli, S. Gasparrini, S. Spinsante, and E. Gambi, "Kinect as a tool for gait analysis: validation of a real-time joint extraction algorithm working in side view", Sensors, no. 15, pp. 1417–1434, 2015.

[15] L. Wiedemann, R. Planinc, and M. Kampel, "Ergonomic-monitoring of office workplaces using Kinect", Proc. Ambient Assisted Living and Daily Activities, December 2014, pp. 275-278, doi: 10.1007/978-3-319-13105-4_40.

Improving the Impact of Wearable Devices in Health Promotion and Wellbeing: the WEHMIX Project

Elena Como
Head of Research and Evaluation
LAMA Agency
e-mail: elena.como@agenzialama.eu

Federico Lucca
Head of Research
Thread Solutions S.R.L.
e-mail: federico.lucca@threadsolutions.com

Abstract— **The promotion of healthy lifestyles has become a priority to tackle the spread of chronic diseases. Digital technologies, such as wearable devices, show great potential to support positive behaviour change and enable healthier behaviours. The WEHMIX project aims to offer new tools and a new user experience to improve the effectiveness and impact of these technologies for health promotion. This paper illustrates the results of the pilot testing of WEHMIX in two separate settings and scenarios of use.**

Keywords-wearable; platform; user experience; impact; digital health.

I. INTRODUCTION

The rise of chronic diseases represents one of the major challenges for health and health care systems worldwide. According to the World Health Organization (WHO), such diseases – in particular cardiovascular diseases, cancers, chronic respiratory diseases, and diabetes – cause the majority of global deaths and burden of disease, affecting in particular the elderly but increasingly also the young and middle-aged individuals [1]. Chronic diseases represent a burden also in economic terms: only in Europe, they absorb around 70-80% of the health care costs, and their negative impact on overall labour supply and productivity has been found to cause a relevant loss of Gross Domestic Product (GDP) and growth [2]. At individual and family level, chronic diseases can have dramatic economic effects, by reducing income and enhancing the costs of necessary care [1].

Fighting the chronic disease epidemic is therefore a policy priority at international level. Fortunately, a key role can be played by prevention, and in particular by primary prevention via the promotion of healthy lifestyles. Indeed, it is widely acknowledged that a considerable share of the global burden of chronic diseases could be reduced or avoided by promoting people's healthier behaviours, in particular by tackling 4 lifestyle-related health risk factors: tobacco consumption, physical inactivity, unhealthy diet, and harmful alcohol use. According to existing research, the adoption of healthier behaviours in these four areas could lead to a reduction of up to 80% the global risk of chronic diseases [3], although other factors as well play a key role, including stress and sleep habits [4].

The WEHMIX for Healthcare project (WEarable Human Machine Interaction user eXperience for Healthcare, WEHMIX in short) aims to contribute to this challenge by proposing a new digital health solution in support of healthier lifestyles. In particular, WEHMIX proposes the integration of multiple wearable devices into a single platform, accompanied by personalised eHealth services to support the adoption and monitoring of healthy lifestyles. This paper presents the results of the pilot testing of WEHMIX in two different settings and scenarios of use. The paper is structured as follows: Section II illustrates the key project assumptions, highlighting in particular the potential of digital and wearable technologies for health promotion and behaviour change. Section III introduces the key innovations of WEHMIX and its distinctive features. Section IV describes the main outcomes from the work in progress, describing the WEHMIX technological platform, its pilot experimentation in the two settings, and the main findings from the user feedback and evaluation exercise. Section V illustrates the main conclusions of the study.

II. MAIN IDEAS, MODELS AND ASSUMPIONS INVOLVED

The use of digital technologies in the field of wellbeing and health promotion is a fast-growing trend. Indeed, there is increasing consensus around the fact that such technologies, and in particular mobile and wearable technologies, can contribute to more personalised prevention, monitoring, and self-management of disease. This consensus is reflected in a number of recent policies at international and national level (e.g., the European eHealth Action Plan 2012-2020) that support the adoption of e- and m-health solutions in various areas including health promotion.

Significant interest is raised in particular by an emergent plethora of wearable devices (smart watches, head/shoe-worn devices, wrist bands, smart shirts, clips, etc.) that are able to measure physical activity levels, as well as number of relevant biometric variables, providing valuable information on an individual's health, at any time and with increasing accuracy. As these devices become smaller, cheaper, and more widespread, interest in them increases; however, it should be noted that the use of such devices cannot be expected to automatically imply better health for their users.

A key question is the following: to what extent can wearable devices become effective and reliable tools for health promotion? To what extent can they have an impact beyond simple monitoring, towards actually promoting actual behaviour change for better lifestyles? A recent review [5] has looked at existing evaluation studies of these devices and provides encouraging evidence of their impact on behaviour change; however, it also underlines that the effectiveness of these devices depends on a number of factors, including their capacity to incorporate adequate incentives and support. Indeed, it is acknowledged that people are empowered, i.e., acquire a higher control of the decisions and actions that affect their health [6], when a number of conditions are met: they are effectively informed about correct lifestyle choices, have a correct perception of their health and of possible risks, are motivated and can see the results of their virtuous choices, are technically and emotionally supported, and can rely on adequate decision support tools to make the best choices for their health [7]. In particular, the possibility to track one's improvements in physical activity seems to motivate people in a constant progress towards their goals, while at the same time it increases the perception of self-efficacy [5]. The user experience (UX) of these wearable devices is another critical factor, since it determines the way in which people "interpret, understand, gain motivation and act on their data" [8], and more generally contributes to determining the long-term retention and motivation to use the device, which has also been found to be a critical issue [9]. All these elements affect the effectiveness of wearable devices for empowerment and behaviour change, and call for their careful design taking into consideration also the human, beyond the technological factors.

The idea of the WEHMIX project is to develop an innovative platform and web app that can improve the user experience and added value of wearable devices for users, hence improving their impact on health and encouraging the change of individual lifestyles for good.

III. INNOVATIVE FEATURES AND POTENTIAL OF THE WEHMIX PLATFORM

The key innovation features of WEHMIX can be summarised as follows.

First of all, WEHMIX allows users to connect different wearable devices to a single digital platform, purposely designed to aggregate the data and to organise it in a smart and easy-to-read way. Thanks to this, users have an immediate and unique access point to consult a synthesis of all their data, without having to access the app of each single device. Moreover, by supporting devices of different brands and types, WEHMIX does not bind the user to a specific brand or device, and if the user decides to change it, WEHMIX ensures perfect integration of the new data into historical records.

Secondly, WEHMIX gives users the possibility to create a personalised health/training plan, with specific goals and activities monitored. In doing so, WEHMIX offers value to a broad range of potential targets: sporty people willing to improve their physical activity, elders willing to monitor their health and prevent chronic diseases, or employees with sedentary jobs willing to improve their wellbeing and productivity thanks to a healthier lifestyle.

Thirdly, WEHMIX allows users to access personalised support from a team of health and wellbeing professionals, in particular for goal setting, data interpretation, and feedback on progress made. In order to connect the user with the professionals, WEHMIX offers an internal chat service; moreover, it provides a specific platform interface for the professionals to visualise in real time the user's data.

IV. MAIN OUTCOMES FROM WORK IN PROGRESS

A. *Technology Development*

The WEHMIX platform is made of three main parts:

- a backend side, connected with devices from major brands
- a mobile app, to allows customers to interact with the data and with the doctor or coach from remote
- a web portal, for the doctor or coach to view the data recorded by the user's devices.

The core of WEHMIX is therefore a *modular* distributed platform that is able to interact with a high number and variety of devices (potentially, any) chosen by the users on the market. A modular model is used by WEHMIX to work with different kinds of data models of different brands and devices, retrieving data from the cloud and giving them back to the user and the professionals in a clear and standardized format. A broad variety of devices has been bought and tested by WEHMIX during past the 2 years, and finally, three devices have been chosen for the first development of the platform, based on an evaluation of quality, price, and errors. These are: Garmin Vivosmart smartwatch for activity and heart rate recording, Fitbit Aria scale for weight and body mass composition, and iHealth for blood pressure measurement.

WEHMIX is based on API LED architecture by the MuleSoft framework MULE ESB, which permits a quick integration with third-party API of different wearable device brands (Figure 1). With MULE ESB, we can expose different layers of API to control the system security. The mobile app and the web app are developed with Ionic and AngularJS framework and communicate with the server side via a JSON REST interface, exposed by the superficial layer of the stack of MULE ESB.

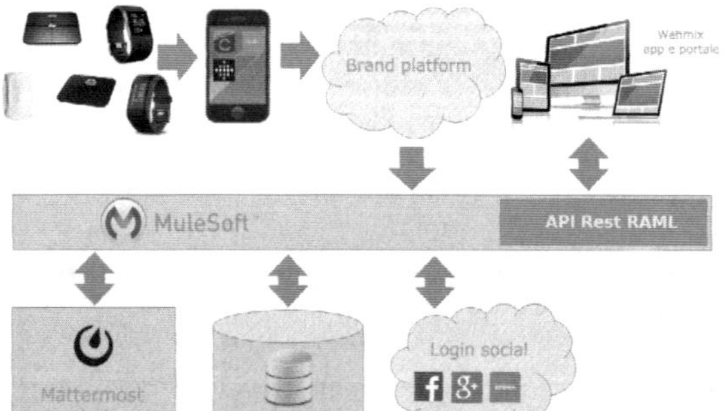

Figure 1. WEHMIX structure

In the eHealth environment, the standard that we decided to use is Open mHealth, promoted by a startup for the integration of all the eHealth services in the world. The idea at the basis of Open mHealth is clearly stated on the company's website: "In healthcare, common data schemas are particularly important because of the semantic importance and complexity of health data. For example, the distinction between fasting and non-fasting blood glucose is critical to its clinical meaning [...] Our common schemas define the meaningful distinctions for each clinical measure, increasing the overall clinical utility of digital health data and improving the ability of developers to quickly build clinically usable products" [10].

Ensuring privacy of the data constitutes one of the main priorities for WEHMIX, and a constant area of research and improvement. After the first release, we moved our core database into CHINO.io, an awarded start-up focused on the privacy of health data, and started working in partnership with them on this project.

B. *Use Cases and Experimentations*

The WEHMIX platform has been pilot tested with 20 people in two separate experimental settings and scenario of use: a Corporate Wellness (CW) scenario and an Active and Healthy Ageing (AHA) scenario.

The CW experimentation was done involving 12 employees of two insurance companies based in Milan. Each user received the monitoring devices (Garmin Vivosmart, Fitbit Aria, iHealth smart scale), was visited by 2 professionals (1 general practitioner and 1 personal trainer), and was given access to the WEHMIX platform and services. To start, each user set his/her own personal plan with the help of the professionals, subsequently they used the devices and platform to monitor their weight, body mass composition, sleep, heart rate, blood pressure, and physical activity, for a period of 3 months. The users could at any time consult their wellness plan and monitor progress via the mobile or web app of WEHMIX; the same data was also visible in real time by the professionals.

The AHA experimentation was developed with a group of 8 users aged between 50 and 67, recruited via a physiotherapy studio located in Cologna Veneta, a town in the Italian province of Verona. They included people with minor and temporary physical issues, which were taken into consideration but not directly addressed by the intervention. In this experimentation, the physiotherapist acted as the supporting professional for the user group, and a slightly different model was tested: the users were given only the Garmin Vivosmart bracelet for tracking physical activity, heart rate, and sleep, while the other two devices (iHealth sphygmomanometer and Fitbit Aria scale) were placed in the studio and used once weekly together with the physiotherapist.

The expected impacts of both experimentations were to raise awareness of users concerning their health and lifestyle, improve attitudes and incentives towards a healthier habit, improve trust in the new technologies for health, improve user experience and usability of the devices, enable the production of data in an easily shareable way, and ultimately improve health, wellbeing, and quality of life and work.

C. *User Feedback and Evaluation*

The two experimentations were used as an opportunity to not only test the technology, but also better understand its potential impacts, and to collect direct and practical feedback for improvement by the users. We carried out the evaluation using a two-step process, consisting of a pre-test analysis (aimed at capturing an initial profile of each user, understanding their previous experience with health and fitness technologies, and collecting their expectations), and an ex-post evaluation (aimed at analysing impacts and collecting actual feedbacks).

We used a qualitative evaluation approach, with a combination of standard questionnaires and in-depth individual interviews; this was considered ideal for the small number of users involved (which allowed for direct contact and more in-depth analysis), and for the type of feedback and indications that we aimed to collect at this stage.

Altogether, the evaluation confirmed the potential of WEHMIX, and the added value brought about by its key innovations. At the same time, the study shed light on the significant differences that exist in practice between the perceptions and preferences of different target groups, and pointed to the need to design specifically tailored solutions for different markets and scenarios of use. Moreover, the study brought the attention to the need to consider pre-existing limitations and barriers to the use of wearable technologies by each type of user, limitations that cannot be addressed directly by WEHMIX, but are nonetheless an important pre-condition for its success.

Below, we summarise the key points emerged from the feedback of the two experimental groups.

Main uses of the WEHMIX platform. The potential use and added value of WEHMIX was perceived differently by the two groups, which reported different priorities and expectations. In the CW scenario, the majority of the users looked at WEHMIX as a valuable tool for health promotion, i.e., for improving their overall lifestyle and everyday habits, with the goal of improving present wellbeing and preventing future diseases. The users of the AHA experimentation, on the other hand, were more sceptical about the possibility to change their behaviours and lifestyles under the stimulus of a digital technology, and were more interested in WEHMIX as a monitoring tool for keeping track of specific, already known, health conditions.

Flexibility towards the use of different devices. The possibility to connect different devices to the same platform, and to change them over time without losing their data, was perceived as positive by the majority of users in both groups. Nonetheless, this feature was perceived as more relevant by the CW group, especially by the most "expert" technology users, who were more likely to buy and change different devices over time. In the AHA group, the possibility to change devices was appreciated in theory, but not immediately perceived as a priority. Some users, however, showed concern regarding the cost of the devices, which suggests that the possibility to choose their own device (eventually opting for a "cheaper" one) could actually make WEHMIX more broadly accessible for them.

Access to professional support and coaching. The incorporation into the platform of a specific, professional coaching service was considered an important added value of

WEHMIX. In the CW group, this was considered the most important feature, and a truly distinctive feature. Users appreciated the initial health assessment and the support received in goal setting received by the professionals, at the same time they gave suggestions to improve the following interaction with them in terms of remote monitoring and periodic feedback. Users also pointed out that they would prefer the support of specific and complementary experts (e.g., nutrition specialist) rather than general practitioners. Similarly, the users of the AHA group argued that WEHMIX would hardly be valuable for them without the presence of a health professional that looks at the data collected by the devices. This was in line also with a generally lower trust of this group in the value of technology "alone" for the purpose of health and wellbeing. They further emphasised the need that the professional be a trusted person, and rejected the idea that he/she may be a "virtual doctor" (e.g., a chatbot), or a person that only operates from remote. The AHA users suggested that WEHMIX should partner with some trusted and recognisable health institution for this service.

User Experience. The interviews also collected feedback and suggestions regarding the user experience of WEHMIX. Part of the feedbacks concerned basic elements, such as the on-boarding process, the connection of the devices, and the functionality and visual features of the app. Other feedback concerned key aspects, such as the visualisation of the data: here the users stressed the importance of finding the right balance between quantity and clarity of the data displayed, and asked for more detail but without replicating the original source. The visualisation of personalised goals and exercises was seen as a critical aspect and some suggestions were given for improvement.

The most important feedback, however, concerned the desired level of interaction of the platform. In this respect, CW users expressed a preference for higher interaction (e.g., welcome messages, targeted alerts, weekly feedback, etc.), while AHA users preferred a lower level of interaction, avoiding push notifications except for emergency situations.

Economic incentives and rewards. A future version of WEHMIX is expected to include additional features, such as incentives (bonus, rewards, etc.) for the achievement of a certain target. Although this is not present as for now, we asked the users a feedback on this idea. The CW group gave positive feedback, expressing preferences for rewards, such as discounts on their health insurance premium, on the purchase of new devices, or on personalised health consultancies. The AHA group, on the other hand, expressed lower interest and in some cases scepticism: they pointed out that for them healthy behaviours are strictly motivated by the desire for good health, and the idea of a platform offering other types of rewards would make them suspicious regarding the actual quality of the services, which would be seen as more commercial and less professional in some way.

Willinness to pay. Lastly, we asked users about their willingness to pay. Most opted for a low amount (below 5 euros/month) for the basic functionalities of WEHMIX; nonetheless, it also emerged that if all the improvements discussed were introduced successfully, users would start comparing WEHMIX with alternative fitness and wellness programmes (e.g., in gyms or specialised centres), and would be available to consider a higher price, given the higher flexibility offered by WEHMIX.

V. CONCLUSION

The WEHMIX project aimed to provide an innovative solution for improving the impact and user experience of wearable devices in the field of health promotion and wellbeing. The main product developed by the project, i.e., a digital platform able to connect with different wearable devices of the user, and to support the provision of personalised services with the contribution of a team of health and wellbeing professionals, has been tested in two different scenarios of use to collect feedback and identify future challenges and possible improvements.

The evaluation performed confirmed the high innovation potential of WEHMIX, while at the same time shed light on the possible areas of improvement and further development. In particular, our findings shed light on the possible challenges and opportunities emerging from the specific targeting of this and similar services towards specific target groups and customer segments.

ACKNOWLEDGMENT

The WEHMIX project received funding by The Autonomous Province of Trento under L.P 6/1999. Partners of the project are Thread Solutions S.R.L., Nous23 S.R.L., LAMA Agency, Fondazione Bruno Kessler, and Net Wise S.R.L.

In addition to the partners, we thank for their participation and contribution to the experimentations: Kirey Group, Zurich Italia, Marco Corti (MD), Federico Frulloni, Pier Alberto Bressan, and all the users that took part and provided valuable feedback.

REFERENCES

[1] R. Busse, M. Blümel, D. Scheller-Kreinsen, and A. Zentner, "Tackling Chronic Disease in Europe. Strategies, interventions and challenges". European Observatory on Health Systems and Policies, Observatory Studies n.20, 2010. [Online]. Available from: http://www.euro.who.int/__data/assets/pdf_file/0008/96632/E93736.pdf [retrieved: January, 2018]

[2] Directorate General Health and Consumers, European Commission, "Reflection process on chronic disease. Final Report.", 2013. [Online]. Available from: https://ec.europa.eu/health//sites/health/files/major_chronic_diseases/docs/reflection_process_cd_final_report_en.pdf [retrieved: January, 2018].

[3] F. Al Maskari, "Lifestyle diseases: An Economic Burden on the Health Services", UN Chronicle Vol. XLVII No. 2, July 2010. [Online]. Available from: https://unchronicle.un.org/article/lifestyle-diseases-economic-burden-health-services [retrieved: January, 2018].

[4] Institute of Medicine (US) Committee on Health and Behavior: Research, Practice, and Policy, "Health and Behavior: The Interplay of Biological, Behavioral, and Societal Influences", 2, "Biobehavioral Factors in Health and Disease", Washington (DC): National Academies Press (US), 2001.

[5] A. N. Sullivan and M. E. Lachman, "Behavior Change with Fitness Technology in Sedentary Adults: A Review of the

Evidence for Increasing Physical Activity", Frontiers in Public Health, vol.4, p. 289, 2017.

[6] World Health Organisation, "Health Promotion Glossary", 1998. [Online]. Available from: http://www.who.int/healthpromotion/about/HPR%20Glossary%201998.pdf?ua=1[retrieved: January, 2018].

[7] S. Cyril, B. J. Smith, and A. M. N. Renzaho, "Systematic Review of Empowerment Measures in Health Promotion", Health Promotion International, vol. 31(4), pp. 809–826, July 2015.

[8] S. Asimakopoulos, G. Asimakopoulos, and F. Spillers., "Motivations and User Engagement in Fitness Tracking: Heuristic for Mobile Healthcare Wearables", Informatics 2017, vol. 4(1), p.5. Multidisciplinary Digital Publishing Institute, 2017.

[9] P. C. Shih, K. Han, E. S. Poole, M.B. Rosson, and J. M. Carroll, "Use and adoption challenges of wearable activity trackers." IConference 2015 Proceedings, pp. 1-12, 2015.

[10] Open m-Health website. [Online.] Available: http://www.openmhealth.org/documentation/#/schema-docs/overview [retrieved: January, 2018].

Different Ways of Engaging the End-Users in mHealth Services

Lessons learned from Swedish case of "Health in Hand" Project

Yan Hu, Sara Eriksén
Department of Creative Technologies
Blekinge Institute of Technology
SE-371 79 Karlskrona, Sweden
Email: {yan.hu, sara.eriksen}@bth.se

Cecilia Fagerström
Blekinge Center of Competence
Blekinge County Council
SE-371 81 Karlskrona, Sweden
Email: cecilia.fagerstrom@ltblekinge.se

Jenny Lundberg, Hanna Tuvesson, Lina Nilsson
Linnaeus University
Vaxjö/Kalmar, Sweden
Email: {jenny.lundberg, hanna.tuvesson, lina.nilsson}@lnu.se

Malin Hofflander
Hofflander Consult AB
SE-294 95 Sölvesborg, Sweden
Email: malin@hofflander.se

***Abstract*— mHealth services are becoming common all over the world. However, the acceptance and use of mHealth services are not always as expected. In this paper, we discuss the lessons learned from our recently finished research project "Health in Hand". Based on the Swedish case study on using mHealth services to support patients with Type 2 Diabetes (T2D) in living a healthy life, we propose strategies for engaging the end-users in mHealth services. These strategies include moving user-driven participation from the margin to the center, building up a professional education team to support both patients and healthcare providers, strengthening user motivation through gamification and supporting the sharing of knowledge within and between different user groups. We believe that these strategies could improve the design of mHealth services through higher user engagement.**

Keywords-mHealth; user engagement; type 2 diabetes; lessons learned; strategies.

I. INTRODUCTION

We have collaborated in an interdisciplinary Indo-Swedish research and development project, "Health in Hand" [1], concerning design of mobile services for health promotion and disease prevention. The Swedish case within the project focused on how to support people with type 2 diabetes (T2D) in living a healthy life through mobile technologies. In this case, a mobile application for communication of blood sugar values, etc., and comprehensible overview of health history, for self-management and a supportive dialog with healthcare providers, was studied. In this paper, we will present lessons learned from the Swedish case study in the project.

There are about 400 000 Swedes, or 3-4% of the population, who have diagnosed T2D. The risk of T2D increases among elderly people. Among those over 75 years old, an estimated 10% or more have T2D [2]. Diabetes is becoming more and more common, not only in Sweden but around the world.

mHealth is already being used to monitor a variety of conditions of chronic disease [3]. The use of mHealth tools itself engages more people to focus on improving their health. Research shows that mobile based interventions with clinical feedback for patients with diabetes improve glycemic control (HbA1c) in the short-term, compared to standard care or other non-mHealth approaches [4].

This paper presents some points on how we should engage the end-users in mHealth services for sustainable use, based on our experiences from the Swedish case study in the "Health in Hand" project. Section 2 provides a brief background of global T2D and global guidelines for managing older people with T2D. Section 3 gives an overview of our experiences from our Swedish case, while in Section 4, we propose some strategies to engage the end-users in mHealth services. The conclusion and future work are presented in the last section.

II. BACKGROUD

A. T2D: A global epidemic in aging populations

T2D is a chronic progressive disease resulting from an imbalance between insulin sensitivity and insulin secretion. It is the most common form of diabetes, responsible for at least 90% of all cases of diabetes [5]. T2D is a lifestyle disease which typically hits people as they grow older. The number of diagnosed cases of T2D has increased rapidly worldwide during recent decades, to an extent that has lead researchers to speak of a global epidemic [6]. In 2015, according to statistics from the International Diabetes Federation (IDF) [5], 1 in 11 adults worldwide had diabetes, and it was estimated that by 2040 this would have reached a level of 1 in 10 adult worldwide living with diabetes [5]. In 2015, approximately 12% of global health expenditure was spent on diabetes. Yet, at that time, IDF also estimated that 1 in 2 adults with diabetes was undiagnosed.

B. *Global Guideline for managing older people with T2D*

In 2013, IDF, recognizing the relative lack of clinical trials and guidelines concerning elderly people with T2D, published a global guideline specifically focused on managing older people with diabetes [7]. The guideline was developed to provide clinicians with recommendations that assist in clinical management of a wide range of older adults, not only those who are relatively well, but also those who, due to frailty, or dementia, or both, are functionally dependent. “While there is increasing recognition that diabetes care for all people should be individualized it is apparent that for many older people with diabetes, care is sub-optimal and often fragmented leaving a substantial portion of adults with unmet clinical and social need.” [7] The guideline takes as its starting point that informal caregivers are often the primary source of everyday advice, emotional support, and practical help for a large number of older people with diabetes, although this is often overlooked by healthcare professionals involved in diabetes care. One of the key principles underpinning the guideline refers to quality use of medicines, including using non-medicine options first if possible, pharmacovigilance, and de-prescribing. Further, the guideline uses three main categories for older adults with diabetes when it comes to determining how to manage their care: those who are independent, those who are dependent, and those who have a life expectancy of less than 1 year and need end of life care.

III. LESSONS LEARNED FROM THE PROJECT

Admittedly, when the “Health in Hand” project started in 2014, we were focused, in the Swedish case study, on developing mobile services for supporting people with T2D in managing their diabetes and developing and maintaining a healthy lifestyle. We aimed to include not only people with T2D but also family members and informal caregivers, as well as healthcare providers, as important actors and participants in the design process. Our aim was to contribute to transforming healthcare delivery through co-construction of mobile health services which focus on factors that support human health and well-being [8] [9], and not only, or primarily, on factors that cause disease (pathogenesis). We were using Participatory Design (PD) as a research method, of which one of the goals was to promote participatory design for mHealth development with the local design of design methods, techniques, and tools.

A. *Challenges facing Swedish diabetes healthcare*

From our case study, we found that there are several challenges faced in Swedish diabetes healthcare. Most of the patients diagnosed with T2D do not measure their blood glucose regularly. Instead, they are called in to the primary care centers or clinics to have it measured once or twice a month. For security reasons, which currently preclude the use of e-mail and mobile messaging, the healthcare providers spend a great deal of time calling patients on the telephone or sending them regular mail. This results in unnecessary visits if the values are normal, while, on the other hand, for patients whose values are abnormal, visits should have occurred earlier. Some patients measure their blood glucose themselves at home, but in many cases there is little or no support for recording and sharing their values with their healthcare providers.

B. *mHealth for people with T2D*

Our study, with its focus on people aged 65 and older who have been diagnosed with T2D, and healthcare teams working with this group of patients, eventually opened our eyes to issues of functional impairments due to aging (such as poor eyesight, stiff fingers, difficulties in remembering passwords or how to do things from one time to the next on a mobile phone), as well as to multiple comorbidity among an aging population. Managing a healthy lifestyle, for many of these people, involved not only managing T2D but also managing other health- and aging-related issues, making the situation in which they were expected to use mHealth services more complex than we had anticipated. Even healthy elderly patients who are diagnosed with T2D but are not simultaneously having to cope with other chronic diseases may have varying needs concerning mHealth services – and these individual needs could vary over time. Basically, the art of engaging end-users in the design, development and sustainable use of mHealth services hinges on identifying these users’ perceived needs and addressing them in ways that can make a difference in everyday life for the individual end-user. End-users bring valuable insights about user needs and use context which can inform not only design and development of mHealth services, but also choices concerning how these services might be introduced and implemented more successfully in a healthcare context.

One of the people with T2D whom we interviewed mentioned that we should perhaps focus on “windows of opportunity” for providing mHealth for people with T2D, rather than clustering them all in one category and trying to reach them all, at all times. “When you are diagnosed with T2D,” he said, “it comes as a shock. For about 6 months, you are shaken and trying to adjust to living with this disease, which can potentially be fatal. During those 6 months, you are especially susceptible to trying to change your lifestyle.” Those 6 months, he suggested, would be the best time to introduce the mHealth services, because that is when the person diagnosed with T2D is looking for a lifeline. This brings us back to trying to develop ways to communicate with and understand the needs of people with different conditions, and to listen to their point of view.

Even with great interest and motivation to use mobile services in home-based T2D healthcare, older people with T2D felt that they were encumbered by a lack of awareness, knowledge and support for self-management. One T2D patient we interviewed said: "For most patients who have T2D, it is not necessary to go for a check-up at the hospital very often, we can check blood glucose at home ourselves if we are taught to do so. In addition, we can measure our blood glucose at home whenever we don't feel well. If I find any values abnormal, I could communicate with the nurses as soon as possible." She suggested that there should be a technical support team with professionals who have knowledge, both of healthcare and diabetes care, and

Information Technology (IT), to educate users before delivery of mHealth services.

C. mHealth for healthcare support team

In Sweden, the healthcare support team for patients with T2D includes healthcare providers, both in hospitals and primary healthcare centers, and informal caregivers, as well as family members. This support team was supposed to cooperate for providing patient-centered healthcare for T2D patients through mHealth services. However, from our study, we found that at least until now, the promotion of mHealth services is still coming from hospital healthcare workers. In our case, doctors or hospital nurses have been given the access right to the data recorded through the mobile application by the patient. In most situations, patients are suggested by the doctors to use the mobile application. The patients therefore get the impression that the healthcare provider should teach and support them when using mHealth services. One doctor in a diabetes center mentioned, "IT has developed so fast that if we ask our patients to try a new technology, like a mobile application, it seems that we also have the responsibility to teach them to use it, but, actually, we are not good at technology either." The lack of knowledge and competence about IT risks reducing the enthusiasm about mHealth services over time.

From the case study, we also found that each healthcare worker works with approximately 3–5 different IT systems in their daily work, and some amount of repeated work is involved. For example, identical information about a patient is updated manually in these different systems. Not only does this cause extra work and frustration, it is also a security risk in that the information may deviate across different systems. The main tool for digital communication between healthcare workers is an internal journal system that serves to share patients' data among different healthcare providers. Due to security and privacy legislation and concerns, not all the data are shared by all the healthcare providers. Although most T2D patients were taken care of by other healthcare providers than hospital staff, such as primary healthcare center staff, or informal care givers, until now, at least in Blekinge, these groups of healthcare providers have no access to the T2D data recorded from mobile applications. These kinds of interoperability problems are another big challenge for mHealth services in Swedish healthcare.

IV. STRATEGIES TO ENGAGE THE END-USERS IN MHEALTH SERVICES

On the one hand, Sweden's population is ageing; in 2040, nearly one in four Swedes will be 65 years or older [10]. On the other hand, Sweden aims to "be best in the world at using the opportunities offered by digitization and eHealth to make it easier for people to achieve good and equal health and welfare, and to develop and strengthen their own resources for increased independence and participation in the life of society [3]". Therefore it is essential to provide social and health care to elderly people through digitization and eHealth. mHealth services, as one of the potentially effective ways, will play a significant role in future elderly healthcare. How to engage the end-users in mHealth services, especially mHealth for older people, for sustainable use, needs to be addressed and further explored.

A. User-driven design method of mHealth services

As we mentioned before, PD was our overarching research method in the Health in Hand project. One of the project goals has been to promote participatory design for mHealth services development. From our study, we found that PD fits quite well with the current patient-centric paradigm which is commonly referred to in nursing and caring for patients with chronic disease, e.g., T2D. PD in a healthcare context concerns not only the technical platform and solutions, but also the work practices in different healthcare organizations. To sustain long-term engagement in a patient-centric mHealth project, user-driven participation should, we argue, be moved from the margin to the center [11]. When the patients, such as older people with T2D, are engaged in co-design mHealth services, it enables them to learn more about their condition and about developing and sustaining a healthy lifestyle despite their chronic illness. It also allows them to have a say in deciding what kind of mHealth support they need and want, in a constructive dialogue with healthcare providers and other involved stakeholders. When designing a new mHealth service, it is often useful to consider whether the design goal can be achieved by fostering creative uses of services they already are familiar with and using. In the project, we could not just simply bring different new mHealth services to be tested and used in a clinical setting at the hospital or a primary healthcare center, as this would be adding to the frustration and heavy work load of the healthcare providers rather than supporting their work [12]. Instead, we chose a mHealth service which was already partly implemented at the hospital as a starting point for further prototyping to explore user needs and how they could be managed with a further enhanced mHealth service. Although we mainly focused on healthcare workers on the T2D team and relatively healthy T2D patients, end-users with special needs should also be taken into account for early design phase participation.

B. Professional education team to support

Lack of IT knowledge of end-users seems to be a great challenge when it comes to user engagement in mHealth currently. Not only elderly people mentioned this, but also healthcare providers were reluctant to engage with mHealth services due to what they perceived as their lack of IT knowledge. Patients expected the healthcare providers to teach them to use the mHealth service, but in most situations, the healthcare providers had neither knowledge nor time to do so. Most of them required more time to learn to use the new technology themselves. From this, we drew the conclusion that there is a need for some kind of a professional education team to educate and support the end-users of mHealth services, especially concerning technical issues, but also concerning content and how to interpret various health indicators etc. Different educational programs should be designed to address different groups of end-users. A support center for mHealth services should also be set up when these services are delivered to large populations.

C. Strengthen end-users' motivation of mHealth services

In the future, concerning design of patient-centric mHealth services, strengthening the patients' motivation is one of the key factors for achieving long-term use of the services. In many cases of chronic disease, patients are suggested to do physical exercise as a main method of controlling their condition. And one of the future needs of elderly care is effective preventive healthcare [10]. Gamification in mHealth services is one approach which is being tested. Mobile games have shown promise of being a successful way of influencing health behaviors recently [13]. A successful mobile game for health, especially for older people, will not only influence the health behaviors, but also support elderly people in having fun by playing games. Normally, the factors affecting users' motivation and engagement in games for entertainment depend on experienced game developers. For those games for health, we still need excellent game developers to design highly attractive games or game elements. Additionally, the input of healthcare providers and behavioral scientists will be essential to delivering an effective mHealth service [13].

D. Knowledge sharing among healthcare providers

Current Swedish eHealth services have, to some extent, achieved sharing of patients' information among healthcare providers. To deliver an effective patient-centric and empathetic healthcare for older people, who often have more than one chronic disease, or complex health conditions, information sharing is not enough. Knowledge sharing, which is about transferring of ideas, experiences, skills, practices etc. among healthcare providers is needed. Knowledge sharing about what kinds of interaction works best in specific situations with a specific person among healthcare providers will help to cope with different special situations. It will not only improve the work practices of healthcare providers [14], but also increase the sustainable use of mHealth services by patients with different special conditions and in specific situations.

V. CONCLUSION

In this paper, we have introduced a Swedish case study from our recently finished research project "Health in Hand", which is focused on transforming healthcare though mHealth services. We have listed several points based on lessons that we have learned from the case study. We conclude that, to promote mHealth services, sustaining end-users' engagement is crucial. Form our experiences, we propose that applying user-driven design methods, having professional education about and support for use of mHealth services, improving users' motivation through games and support for sharing knowledge among healthcare providers could make some contribution to high end-user engagement in mHealth services in the future.

Digitalization of healthcare needs to be supported on all levels of health care organizations. There is also a need to reserve time and space for testing new ways of working. We have several projects in the planning stage on end-user engagement in mHealth services. In the future, the suggested strategies presented in this paper will be drawn up and further explored within these projects.

ACKNOWLEDGMENT

We thank all the healthcare professionals and staff at the Blekinge County hospital and the primary care centers, the Blekinge Diabetes Association and the people with T2D who took part in the Swedish case study of the Health in Hand project. We also gratefully acknowledge the grant from the Swedish Innovation Agency VINNOVA (#2013-04660).

REFERENCES

[1] S. Eriksén, M. Georgsson, M. Hofflander, L. Nilsson, and J. Lundberg, "Health in hand: Putting mHealth design in context," in *2014 IEEE 2nd International Workshop on Usability and Accessibility Focused Requirements Engineering (UsARE)*, 2014, pp. 36–39.

[2] "Type 2-diabetes," *Diabetes.se*. [Online]. Available: /diabetes/lar-om-diabetes/typer/typ-2/. [Accessed: 22-Feb-2018].

[3] "Vision for eHealth 2025," Ministry of Health and Social Affairs, Sweden, Mar. 2016.

[4] S. Kitsiou, G. Paré, M. Jaana, and B. Gerber, "Effectiveness of mHealth interventions for patients with diabetes: An overview of systematic reviews," *PLOS ONE*, vol. 12, no. 3, p. e0173160, Mar. 2017.

[5] "International Diabetes Federation - What is diabetes." [Online]. Available: https://www.idf.org/about-diabetes/what-is-diabetes.html. [Accessed: 22-Feb-2018].

[6] W. H. Herman and P. Zimmet, "Type 2 diabetes: an epidemic requiring global attention and urgent action," *Diabetes Care*, vol. 35, no. 5, pp. 943–944, 2012.

[7] "International Diabetes Federation. IDF Global Guideline for Managing Older People with Type 2 Diabetes." [Online]. Available: https://www.idf.org/e-library/guidelines/78-global-guideline-for-managing-older-people-with-type-2-diabetes.html. [Accessed: 22-Feb-2018].

[8] B. Lindström, "Salutogenesis – an introduction." [Online]. Available: http://www.centrelearoback.org/assets/PDF/04_activites/clr-GCPB121122-Lindstom_pub_introsalutogenesis.pdf. [Accessed: 22-Feb-2018].

[9] A. J. Mahmud, E. Olander, S. Eriksén, and B. J. Haglund, "Health communication in primary health care-A case study of ICT development for health promotion," *BMC Med. Inform. Decis. Mak.*, vol. 13, no. 1, p. 17, 2013.

[10] "Elderly care in Sweden," *sweden.se*, 14-Jun-2013. [Online]. Available: https://sweden.se/society/elderly-care-in-sweden/. [Accessed: 14-Feb-2018].

[11] V. Righi, S. Sayago, A. Rosales, S. M. Ferreira, and J. Blat, "Co-designing with a community of older learners for over 10 years by moving user-driven participation from the margin to the centre," *CoDesign*, vol. 14, no. 1, pp. 32–44, Jan. 2018.

[12] S. Eriksén, M. Hofflander, L. Nilsson, C. Borg, M. Georgsson, and J. Lundberg, "Transforming Healthcare Delivery: ICT Design for Self-Care of Type 2 Diabetes," *Des. Self-Care Everyday Life*, p. 17, 2014.

[13] D. King, F. Greaves, C. Exeter, and A. Darzi, "'Gamification': Influencing health behaviours with games," *J. R. Soc. Med.*, vol. 106, no. 3, pp. 76–78, Mar. 2013.

[14] S. S. R. Abidi, "Healthcare knowledge sharing: purpose, practices, and prospects," in *Healthcare Knowledge Management*, Springer, 2007, pp. 67–86.

Region-Based Bed Capacity mHealth Application for Emergency Medical Services: Saudi Arabia Case Study

Khulood Alghamdi
King Saud University Medical City
King Saud University (KSU)
Riyadh,
KSA
e-mail: kmalghamdi@ksu.edu.sa

Shada Alsalamah
Information Systems Department
College of Computer and Information Sciences, KSU
Riyadh, KSA. e-mail: saalsalamah@ksu.edu.sa
Media Lab, Massachusetts Institute of Technology,
Cambridge, MA, U.S. e-mail: shada@mit.edu

Ghada Al-Hudhud
Information Technology Department
College of Computer and Information Sciences, KSU
Riyadh, KSA. e-mail: galhudhud@ksu.edu.sa

Thamer Nouh
Trauma and Acute Care Surgery Unit, Department of Surgery, College of Medicine, KSU
Riyadh, KSA. e-mail: tnouh@ksu.edu.sa

Ibrahim Alyahya
Emergency Medical Services
Saudi Red Crescent Authority
Riyadh, KSA. e-mail: imyahya@srca.org.sa

Sakher AlQahtani
Department of Pediatric Dentistry and Orthodontics
College of Dentistry, KSU
Riyadh, KSA. e-mail: asakher@ksu.edu.sa

Abstract—**One of the major causes of death in trauma cases is ambulance diversion due to the unavailability of resources (bed and specialists) in the closest hospital. To date, there are no solutions in the literature that link Emergency Medical Services (EMS) with trauma centers in Saudi Arabia to avoid Emergency Departments (EDs) crowding and shortages in terms of inpatient bed capacity. Therefore, this study aims to bridge the gap by helping paramedics deliver patients to the nearest trauma center with available resources in the shortest possible time, as well as to monitor ambulance diversion data. This is achieved by regionalizing the availability of resources in the nearest hospitals using a mobile health (mHealth) application and two algorithms: Dijkstra and Fusion. The former helps route the ambulance's path taken, while the latter predicts inpatient bed capacity. The solution is an effective and time-optimizing application for the rescuer or paramedic, ED Support Officer (EDSO), and EMS administrator. It should help optimize EMS resources, reduce ED crowding, and increase the quality of urgent care services.**

Keywords- Emergency medical services; Operating room; Emergency department; Electronic health record; mHealth.

I. INTRODUCTION

Ambulance diversion is a strategy often used by overcrowded hospitals and Emergency Departments (EDs) with unavailable beds or resources. EDs that do not have available inpatient beds for emergency patients due to inefficient processes or environmental disaster, often have to divert ambulances to another hospital. In this case, the targeted hospital notifies the ambulance to go to another ED at any other hospital. This increases the total time it takes the ambulance to transport the patient and, thus, can have negative consequences on the patient's health and safety, especially in cases of urgent care when every second is crucial and could mean life or death. This is evidenced in trauma treatment guidelines which state that to reduce risks, it is strictly suggested that the total time should be under 60 min for trauma victims. In research published in the Journal of the American Medical Association [1], researchers found that victims of a heart attack are more likely to die when their nearest hospital had high diversion rates of more than 12 hours diversion per day in contrast to other heart-attack victims who happened to be near a hospital and without any need for diversion [1]. Therefore, to eliminate the need for ambulance diversion and reduce the time it takes to reach the nearest available emergency department, there is an urgent need for a continuous, stable process that connects both the ED and the ambulance services together.

Furthermore, an arriving patient may be declined or placed on hold by Medical Care Utilization (MCU), if there is no bed available to accommodate the patient in the unit. In such cases, the patient may be subject to similar health risks due to the necessity of finding an alternative health care provider or waiting until a bed becomes available. From the hospital's perspective, the potential revenue is either lost or delayed, which may lead to some financial inefficiencies. Similarly, an arriving patient may be accepted for treatment but can be accommodated in another MCU (e.g., an arriving obstetrics patient can be boarded in neurosurgery), and hence there is the chance that it may not be possible to move admitted patients from the ED due to an inpatient bed problem. This forces the ED to board admitted patients until inpatient beds are available, effectively reducing the ED's capacity to care for new patients. Boarding of inpatients in the ED has also been cited as the most important determinant of ambulance diversion.

In addition to time and systems integration factors, cyber attacks on healthcare infrastructure for a deliberate interruption of healthcare services could leave hospitals no choice but to divert ambulances. Global attacks on electronic service providers in healthcare, business, and government sectors have been witnessed in recent times and there has been a wave of deliberate attacks using WannaCry Virus. WannaCry is a malware of Trojan virus called Ransomware. The virus holds the infected computer hostage and demands that the victim pays a ransom to regain access to the files on his/her computer. Ransomware, like WannaCry, works by encrypting most or even all of the files on a user's computer. Hence, the software demands that a ransom be paid in order to have the files decrypted. The episodes of Ransomeware attacks [2] have had a massive scale effect on around 150 countries, targeting not only home computers but also healthcare, communications infrastructure, logistics, and government entities for financial gain. This challenged healthcare providers' deployment of their computerized complex systems to maintain and keep patients' records and patient diversion data at critical times and so paralyzing healthcare services totally in some targeted health trusts in many countries, such as the British's National Health Service (NHS) [2]. Hospitals across England and Scotland's NHS reported that the cyber attack was causing disruptions to their health services by "affecting X-ray imaging systems, pathology test results, phone systems and patient administration systems" [2].

This project aims to address ambulance diversion issues in Emergency Medical Services (EMS) and monitors ambulance diversion data. Therefore, it focuses on finding a practical solution to facilitate the delivery of EMS across all EDs without ambulance diversion to reduce the time it takes them to reach the nearest ED with the available inpatient beds and the right resources for treating the patient. The remainder of this paper is organized as follows. Section II describes the background, while Section III discusses related work. Section IV presents our methodology, and conclusions are drawn in Section V.

II. BACKGROUND

The Saudi Arabian government has accorded high priority to healthcare services. According to the Basic Law of Saudi Arabia Rights of the Saudi Citizen, the government guarantees the right to healthcare for citizens and their families and the government is responsible for providing public health care services to all Saudi citizens [3]. In recent years, healthcare services in Saudi Arabia have improved tremendously in terms of both quantity and quality, and this is evident in the literature and government white papers in general, and reflected in the total number of hospitals in Saudi Arabia in particular. Back in 2015, there were 462 hospitals in Saudi Arabia with a capacity of 69,394 beds, with an increment of 49 hospitals in comparison with their number in 2011. The governmental sector owns nearly 70 percent of hospitals and the rest are owned by the private sector [4].

A. ED Crowding

ED crowding and inpatient bed capacity are pressing problems facing healthcare worldwide. ED Crowding is defined as:

> "*A situation in which the ED function is impeded by the number of patients waiting to be seen, undergoing assessment and treatment, or waiting for departure, exceeding the physical or staffing capacity of the department.*" [5]

Instead of managing ED crowding as a standalone problem in order to eliminate it, contributing factors must also be examined. First, patients' long wait until an inpatient bed becomes available and the increased inpatient waiting time for care are also indications of ED crowding. Second, the lack of hospital inpatients bed capacity could lead to ED boarding, which is a significant cause and one of the main reasons for ED crowding [6]. ED boarding is the practice of keeping patients in the ED waiting area due to the lack of available inpatient beds, even after their admission to the hospital. This results in many issues, including ambulance diversion, extended patient waiting periods, delays in treatment and longer waiting times for other patients who do not require admission to be treated [7]. Finally, there is an urgent need for a solution to eliminate ED crowding and ambulance diversions resulting from unavailable inpatients bed capacity and ED boarding. This is to provide emergency patients with speedy and reliable healthcare and reduce the time it takes the ambulance to reach the nearest available emergency department. To address the above issues, there is a need for a continuous, stable process that requires a system-wide support among all healthcare related parties, in order to connect and work with both the emergency departments and the ambulance services together.

B. Hospital Bed Capacity Planning

Hospital bed capacity planning is crucial in healthcare because it is essential for managing hospital resources and hospital staff and personnel. In addition, it could be the deciding factor between a patient's life and death. In some countries, such as Finland, New Zealand and Germany, the unit for measuring hospital care and its capacity is bed occupancy rate [9], which is defined as "*the number of hospital beds occupied by patients expressed as a percentage of the total beds available in the hospital*" [8]. This rate remains an essential unit in hospital capacity planning. Nevertheless, using bed numbers and occupancy as a measurement in hospital capacity planning will not foretell the hospital's future demand; neither will it provide a valid estimate of hospital services [9].

In research published by the World Health Organization, the researchers propose using strategies that focus on the benefits of using systematic processes in hospital capacity planning. They argue that it is not beneficial to look at the hospital from the perspective of beds and its occupancy rates, but rather it is necessary to focus on processes and the path taken by the patients inside the hospital. One of the strategies mentioned is to design hospital flows around "care pathways" instead of counting beds; the strategy works by identifying the variety of pathways the patients take inside the hospital, as well as the factors that can cause delays in patients' treatment and this could be displayed in the form of a bottleneck [10].

Therefore, the key to successful capacity planning is to try to eliminate any possible future cause of a bottleneck; sometimes this could be the number of inpatients' available beds, ineffective allocation of existing patients among different medical service units and sometimes it could be other hospital departments attempting to enhance their performance without realizing how this might affect others. Guaranteeing that there are as few bottlenecks as possible will, in turn, result in minimizing the delay in patients' treatment, separating patients into two streams based on complexity rather than urgency, and creating a fast-track patient stream for patients who can be treated and discharged more or less immediately [9].

C. *International Hospital Statistics*

Despite hospital planning strategies, this section highlights some international statistics for acute hospital bed shortages around the globe [11], along with a definition of hospital beds in such countries [11] (summarized below in Table I). In Austria, hospital beds have an average length of stay of 18 days or less. This includes some day care beds. In Germany, acute hospital beds are those other than psychiatric and long-term beds, and that exclude any day care beds. In Iceland, acute hospital beds are calculated from bed-days, assuming 90% occupancy rate; beds in medicine and surgeries of main hospitals and mixed facilities are available in small hospitals that do not include any day care beds. In Italy, acute hospital beds include in-patient beds of psychiatric hospitals and in-patient beds of psychiatric wards of other hospitals; these do not include any day care beds. In Spain, acute hospital beds include general hospitals, maternity, other specialized hospitals, and health centers; no day care beds. Last but not least, in the UK, acute hospital beds include NHS acute medical, surgical, and maternity beds (excluding Northern Ireland).

TABLE I. DEFINITIONS OF HOSPITAL BEDS IN SELECTED COUNTRIES [11]

Country	Content
Austria	Beds in hospitals with average length of stay of 18 days or less
Germany	Beds other than psychiatric and long term beds
Italy	In-patient beds of psychiatric hospitals and in-patient
United Kingdom	National Health Service acute medical, surgical and maternity beds (excluding Northern Ireland)
Spain	General hospitals, maternity, other specialized hospitals, health centers
Sweden	Beds for short term care run by county councils and 3 independent communities (short-term includes medical, surgical, miscellaneous medicine/surgery, admission department and intensive care)
Turkey	Public hospitals, health centers, maternity hospitals, cardiovascular and thoracic surgical centers orthopaedic surgery hospitals

III. LITERATURE

A. *Bed Capacity Planning Approach*

Healthcare, like any other industry, faces huge pressure to improve efficiency and reduce costs. A study by McKinsey [5] points out that the U.S. spends at least $600 - $850 billion on healthcare annually. One area that can be leveraged in healthcare is the support of informed decision-making processes. This is to allow the end user (namely, hospital administrators or clinical managers) to assess the efficiency of existing healthcare delivery systems.

Discrete-Event Simulation (DES) is a widely used technique for the analysis of systems with complex behaviors [13]. DES has been widely applied in healthcare services [13] to study the interrelationships between admission rates, hospital occupancy, and several different policies for allocating beds to MCUs. Lewis [14] studied bed management in Germany's hospitals, and decision support systems were presented depending on mathematical approaches and computer-based assistance designed to improve efficiency and effectiveness of admission planning and bed assignment. The study interviewed professionals in bed management to identify aspects that must be respected when developing the decision system, ensuring that patients' treatment priorities and individual preferences are respected [14]. Patient admission and assignment is based on up-to-date and flexible length of stay estimates, being taken into aggregated contingents of hospital beds, treatment priorities, patient preferences, and a linkage between clinics and wards [14].

The main reason for using DES for modeling a healthcare clinic instead of other mathematical modeling tools (such as linear programming and Markov chain analysis) is the ability to simulate complex patient flows through healthcare clinics, and to play "what if" games by changing the patient flow rules and policies [14]. Such flows are usually in emergency rooms, where patients can be seen without appointments, and require treatment for various sets of ailments and conditions [14]. These disorders can range from mild injuries to serious medical emergencies. Although the number of patients is unpredictable, medical staff can control the treatment by minimizing patient waiting times and increasing staff utilization rates [15].

In emergency rooms, to reduce waiting times of low priority patients, Schmidt et al. [15] analyzed the effects of using a fast track lane. As emergency rooms are prioritized according to the level of patient sickness, low priority patients may have to wait for exceedingly long periods of time [15]. A simulation model was used to classify daily occupancy distributions; it helps in studying the swapping of overflow and bed capacity levels, and it investigates the effects of various changes. ED overcrowding principally results from the incapability of admitted patients being transferred to ward beds in a timely manner [19]. Most experts agree that a greater inpatient capacity is required in order to relieve access block and decrease ED overcrowding

[15]. Schmidt et al. collected real data from a single month at a single hospital, and a computer model was developed to examine the relationship between admissions, discharges and ED overcrowding (the number of hours admitted patients waited in ED before transfer to an inpatient bed). Meanwhile, Schmidt et al. proposed in [15] a facility location model to locate ER services on a network and determine their respective capacity levels, such that the probability of diverting patients is not larger than a particular threshold.

Lin [16] presents a conceptual model of ED overcrowding to help administrators, researchers, and policymakers. The ED conceptual model recognizes at least three general categories of care delivered in the ED: emergency care, unscheduled urgent care, and safety net care. The outputs are patients who are unable to obtain follow-up care and often return to the ED if their condition does not improve or deteriorates [16]. The throughput component of the model identifies patient length of stay in the ED as a potential contributing factor to ED crowding. The ED crowding is then measured based on two phases; the first includes triage, room placement, and the initial provider evaluation. While the second phase of the throughput component includes diagnostic testing and ED treatment. The triage phase is used to objectively identify patients suitable for treatment by emergency nurse practitioners [16]. Since emergency nurse practitioners show high diagnostic accuracy, the emergency nurse practitioner model of care is considered an important strategy in reducing the length of stay of ED patients and may prevent ED crowding [16].

The input-throughput-output conceptual model of ED crowding may be useful for organizing research, policy, and operations management agenda to alleviate the problem [16]. This model illustrates the need for a systems approach with integrated, rather than piecemeal, solutions for ED crowding [16]. In the study, there are four general areas of ED crowding that require future research. First, research must consider developing valid and reliable measures of ED crowding. These measures should be sensitive to changes throughout time. Second, research in the field should identify the most important causes of ED crowding from each component of the model. Third, the effect of ED crowding on the quality of patient care must be assessed. Finally, interventions to reduce ED crowding need to be evaluated.

Infective allocation of existing bed capacity among different medical service units can lead to service quality problems for the patients along with operational and/ or financial inefficiencies for the hospitals [16]. Most health care managers apply relatively simple approaches, such as the use of target occupancy level with average length of stay, to forecast bed capacity required for a hospital or an MCU [16]. Yet, the failure to adequately consider uncertainties associated with patient arrivals and the time needed to treat patients by using such simple approaches may result in bed capacity configurations where a large proportion of patients may have to be turned away. The application of queuing theory allows for the evaluation of the expected (long-run) performance measure of a system by solving the associated set of flow balance equations [16].

Other research considers the hierarchical relation between care units. For example, after a mother-to-be delivers her child in the labor and delivery unit, she should be moved to the postpartum unit for recovery [16]. If the ability downstream is insufficient, patients must stay within the current care units with typically more costly equipment, thereby meaning the full capacity is reached at these upstream care units [16]. To take the interactions among care units in a hospital into account, Lin [16] first applies queuing network methodology (without blocking) to discover a balanced bed allotment, which is obtained through trial-and-error work. Then, Lin uses simulation analysis to estimate the blocking behavior and patient sojourn times. The authors [16] develop a mathematical programming formulation to address this problem and the system uses a different approach by integrating results from queuing theory into an optimization framework. Specifically, the model with each MCU in a hospital has an M/M/c/c queuing system to estimate the probability of rejection when there are beds.

B. International Bed Management System (BMS)

Hospitals should use BMS that provides a real-time display of hospital occupied beds, along with the available beds, as well as the current status of each one [16]. Therefore, by using a BMS system, the hospital staff would be able to view the status of each bed, whether the bed was occupied, vacant or being prepared for a patient. Also, the hospital staff would be able to view the patient's status; if the patient was going to be discharged, had already left, or was being transferred. Moreover, by using the system, nurses can utilize more of their time in caring for the patient, instead of handling bed assignment tasks manually. This Bed Management Unit used at Alexandra Hospital provides critical benefits to both patients and staff; indeed, by using the system, patients' waiting times have been decreased by 30 percent [17]. Singapore General Hospital (SGH) is another hospital that has benefited from the use of technology to manage hospital beds. The SGH uses BMS technology to help improve hospital capacity and care. BMS is a web-based system that allows hospital staff and administration to access information related to patient flow anywhere in the hospital. The BMS user interface has been configured to display the location of patients in the hospital as well as the primary physician. The system also allows the hospital staff to view and track information, records and any specific actions related to the patient's needs, and any follow-up movements required. The system gives nurses and bed-management staff a full overview of the bed status and patient needs, which allows them to take action immediately.

The way this system works is as follows: when the patient is admitted to the hospital, they receive a Radio

Frequency Identification (RFID) tag with a unique identifier that will identify and track the patient's location during their stay at the hospital. The system then searches for a bed that best fits the patient's needs, based on their condition before assigning that bed to the patient. The system also uses a real-time location system to help identify the location of the patient through the tag and display; this creates a workflow related to the patient's movements. Hospital departments are also provided with Liquid Crystal Display (LCD) panels that display BMS dashboard with real-time patient RFID location, which shows the patient's' information and bed statuses automatically in real-time. Once the patient is discharged, the BMS system will notify the housekeeping staff through their Personal Digital Assistant (PDA) to clean and prepare the vacated bed, and once the bed is ready, the housekeeping staff updates the bed status [18].

C. Nature of Trauma

In physical medicine, major trauma is an injury or a damage to a biological organism caused by physical harm from an external source [28]. Major trauma is also an injury that can potentially lead to serious long-term outcomes, such as chronic pain or other lifelong ailments. There are different types of trauma [28]:

1) Birth trauma: an injury to the infant during the process of being born. In some psychiatric theories, the psychic shock is produced in an infant by the experience of being born [28].

2) Psychic trauma: a psychologically upsetting experience that produces an emotional or mental disorder or otherwise has lasting negative effects on a person's thoughts, feelings, or behavior [28].

3) Risk for trauma: a nursing diagnosis accepted by the North American Nursing Diagnosis Association, defined as accentuated risk of accidental tissue injury, such as a wound, burn, or fracture.

The initial evaluation of a trauma patient is a challenging task and time-critical as every minute could be the difference between life and death. Over the past 50 years, the assessment of trauma patients has evolved because of an improved understanding of the distribution of mortality and the mechanisms that contribute to morbidity and mortality in trauma [27]. On the one hand, early deaths may occur in the minutes or hours after the injury. These patients frequently arrive at a hospital before death, which usually occurs because of hemorrhage and cardiovascular collapse [27]. On the other hand, late trauma mortality peaks in the days and weeks after the injury and is primarily due to sepsis and multiple organ failure [27]. Therefore, systems supporting trauma care focus on the treatment of a patient from early trauma mortality, whereas critical care is designed to prevent late trauma mortality. This is the reason why it is important to find beds prior to trauma cases [28], and therefore, our proposal strives to address this matter.

D. Hospital Bed Capacity Globally and Saudi Vision 2030

Hospitals' provision for accommodating the increasing numbers of emergency admissions is a matter of considerable public and political concern and has been the subject of widespread debate [25]. When discussing a hospital's bed capacity, a number of questions are often raised. First, what is a hospital bed? A bed is merely an item of furniture on which a patient can lie. For a bed to make any meaningful contribution to a healthcare facility's ability to treat someone, it must be accompanied by an appropriate hospital infrastructure, including trained professional and managerial staff, equipment and pharmaceuticals [12].

For several years, hospital managers have been under pressure to reduce bed capacity and increase occupancy rates for operational efficiency, especially in the Hajj season (i.e. Annual Islamic pilgrimage to Mecca, Saudi Arabia) [25]. More recently, public concern has arisen in cases where patients could not gain access to a local hospital or were subjected to extended delays for the availability of vacant beds [20]. Many countries now struggle to provide cost-effective, quality healthcare services to their citizens [23]. Saudi Arabia has experienced high costs along with concerns about the quality of care in its public facilities [26]. To address these issues, Saudi Arabia is currently restructuring its healthcare system to privatize public hospitals and introduce insurance coverage for both its citizens and foreign workers [26]. These changes provide an interesting and insightful case for the challenges faced when radically changing a country's healthcare system. The situation also demonstrates a unique case in the Middle East for greater reliance on the private sector to address rapidly escalating healthcare costs and deteriorating quality of care [26]. The complexity of changing a healthcare system is discussed with the many challenges associated with the change.

According to Saudi Vision 2030 [22], the healthcare system has benefited from substantive investment in recent decades. As a result, we now have 2.2 hospital beds for every 1,000 people, world-class medical specialists and an average life expectancy rising from 66 years to 74 years in the past three decades [24]. Work is currently underway to build and develop 38 news hospitals with a total capacity of 9,100 beds, in addition to two medical sites accommodating 2,350 beds [24]. During the current fiscal year, 1437/1438, 23 new hospitals (4,250 beds) in various regions across Saudi Arabia were built [24].

E. Availability of Beds in Saudi Hospital ED

Implementing a Saudi Arabia-wide system would allow a patient's referral from one healthcare provider/ facility to another. This includes the ability to electronically transfer patient-related data in either a structured (namely, organized and well-maintained information that can be obtained in a simple click), or non-structured fashion (namely, data which is energy consuming and hard to handle data). Alternatively, pointers to electronic health accessible data could be used, including patient diagnosis and treatment, referral notes, medication lists, laboratory test results, radiology reports, digital images, audio and video files. This solution would enable integration of information on the availability of the facility, bed, provider or specialty. In addition, such a solution supports optimizing the search for best-fit resource utilization. The proposed work supports:

- Riyadh hospital's bed management program with automation, including integrated interfaces with Hospital Information System (HIS) *"as an element of health informatics that focuses mainly on the administrational needs of hospitals."*
- Centralized query capabilities for both head quarters and regional administrators.
- Operational support to hospitals and primary health care practitioners providing patient referrals.
- Support full inpatient bed management cycle-interface with multiple systems, including registries, HIS, and communication systems. Generate messages to hospital housekeeping.
- Emergency bed requirements and other hospital departments across Saudi Arabia's regions to inform of status - full reporting and analytical capability [24] (See Table II below).

TABLE II. HOSPITALS ESTABLISHED IN SAUDI ARABIA (2010-2013) [24]

Regions	No. of Hospitals Established	No. of Beds
Riyadh	8	1,400
Makkah	8	2,386
Eastern	7	1,150
Al Madinah	5	650
Hail	3	180
Qassim	4	475
Northern Border	3	400
Asir	8	800
Tabuk	3	350
Jouf	5	550
Baha	2	100
Jazan	6	400
Najran	3	150
New	45	6,521
Upgraded/Replacement	65	2,470
Total	**175**	**17,982**

F. Regionalization Vs. Bed Capacity

In a healthcare service facility, when an ER is full or all intensive care beds are occupied, hospitals send out a divert status. When a hospital is on divert status, incoming patients might be sent to hospitals which are further away or kept at the hospitals where they currently are that yet may not be able to provide adequate service. For a critical trauma victim, the consequence of divert status can be the difference between life and death. Therefore, the healthcare service facility needs to construct a facility location model that can simultaneously determine the number of facilities opened, their particular locations, and their capacity levels. This should ensure that the probability of all servers in a facility gets busy does not exceed a pre-determined level.

To achieve such capabilities in a facility location model, some solutions in the literature incorporate queuing systems into facility location models to consider the chance of servers' availability and focus on reducing the demand lost due to the shortage of capacity or system congestion. However, to date there are no available region-based bed-capacity systems in the literature for Saudi Arabia. Therefore, this project aims to bridge this gap by locating ER services across Riyadh City on a network and determine their respective capacity levels such that the probability of diverting patients is not larger than a particular threshold. This proposal should enhance the performance of EMS in Saudi Arabia in general, and Riyadh City in particular. In Addition, it should reduce the total number of deaths resulting from trauma diversion.

IV. METHODOLOGY

This research uses a mixture of qualitative research methods for data collection and system design. First, a total of 34 hours of semi-structured interviews with 23 multi-disciplinary experts were conducted. The sample of interviewees included 1 trauma surgeon, 1 physician, 2 health informatics experts, 4 information security experts, 8 computer scientists, and 7 Saudi Red Crescent Authority (SRCA) representatives with information technology and medical backgrounds, including the General Manager of Emergency Medical Services, SRCA. Through those interviews, primary data was collected about all trauma centers in Riyadh, the number of states in emergency, the data assessment sheet EMS personnel uses today (in paper format) that needs to be sent to hospitals with notification along with patient ID number, and the challenges faced by the ED. Second, soft systems methodology was used to design the proposed system and produce an architectural design (See Figure 1).

A. Region-Based Bed-Capacity System Framework

The SRCA EMS personnel interviews (mainly with the Business Analyst [30]) provided substantial evidence that hospital beds can reduce deaths during emergencies and delay treatment of many patients in emergency by devolving the bed capacity system. Therefore, this research aims to devolve web services (See Figure 1) that connect the proposed mHealth application with any Electronic Health Record (EHR) in any hospital to show the bed capacity. Furthermore, to learn more about ED process, we interviewed Dr. Thamer Nouh [31], Trauma Surgeon at King Khalid University Hospital, to identify the criteria relating to emergency department bed capacity in trauma care in one of the large tertiary hospitals in Riyadh (namely King Khalid University Hospital).

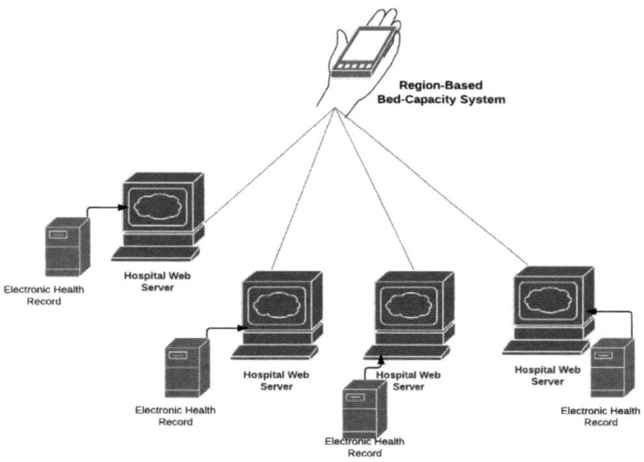

Figure 1. Proposed System Overview.

B. System Functional Requirements

Based on two sets of interviews the following solution is proposed:

- The EMS rescuer/paramedic can view information about hospitals across the city of Riyadh.
- The information available in the application must be dynamic and in real-time, which means that the paramedic can see the available and current information about the hospitals at any time and anywhere across Riyadh.
- The EMS rescuer/paramedic can view each hospital's exact location, bed capacity and the available medical resources to suit a patient's case.
- The EMS paramedic can view available inpatient bed capacity in real-time for the hospitals.
- The EMS paramedic can find the nearest suitable hospital location for a patient's case with the right medical services for this patient's emergency situation, to guarantee a transfer in a speedy manner that would prevent negative implications to the patient and, at the same time, provide the right health care at the right time in a speedy manner.
- The Emergency Department Support Officer (EDSO) [32] receives notification of the new patient and can view their assessment information.
- The administrator of EMS can view reports about every EMS rescuer/paramedic case and patient information.

C. Workflow Design

Based on the collected data, four criteria are used to decide on bed availability:

- Available bed in radiology department,
- Available bed in Intensive Care Unit (ICU)
- Available bed for inpatient
- Available bed in Operation Room (OR)

A final criterion would be existing specialists in one of the three fields (Orthopedic surgery, Neuro surgery, and Emergency surgery). However, the EMS cannot decide which patient needs to go to the radiology department for examination unless the specialist visits the radiology department. Hence, the system will be in green mode, then the EMS rescuer/paramedic can choose a suitable hospital. Accordingly, ER can be available if there are enough beds in the radiology department, ICU, inpatient and OR (See in Figure 2). The system has two components:

- In the first part of the system, an attempt is made to construct module web services integration with each internal hospital system to decide ability of ER to receive the new patient; this depends on the criteria that are mentioned above.
- On the other hand, EMS regenerates bed capacity of Riyadh hospitals in trauma cases to decide on a suitable hospital, depending on the following:
  - Shortest way from the hospital location that guarantees a speedy transfer that would prevent negative implications.
  - Hospitals have the medical resources to treat the patient's case.
  - Hospitals have specialized personnel to treat the patient's case.

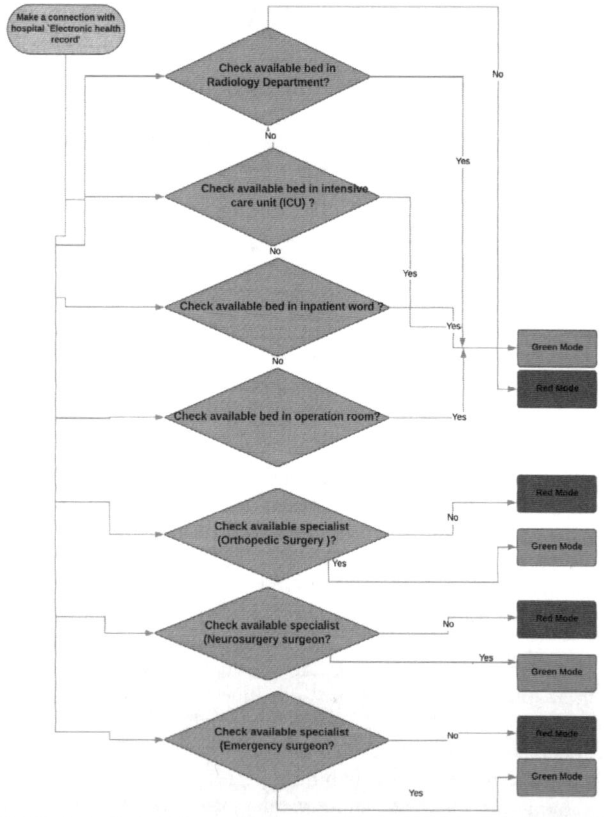

Figure 2. Proposed System workflow.

D. Process Algorithms

The goal of this phase is to build a decision support model that is powerful, robust, comprehensible, optimal, and effective. There are many different search algorithms that can run the best diction support model (for example, Fusion [33] and Dijkstra's [34]). The best algorithm to be chosen depends on the collected data.

On the one hand, Fusion Algorithm [33] is used in many tracking and surveillance systems. One method for the design of such systems is to employ a number of sensors (perhaps of different types) and to fuse the information obtained from all these sensors on a central processor. Past efforts to solve this problem required the organization of feedback from the central processor to local processor units. This can be used by the hospitals for proper surveillance of accidents on the roads. It can also reduce the time necessary for bringing the patient to the hospital [32].

On the other hand, Dijkstra's Algorithm [34] finds the shortest path from a starting node to a target node in a weighted graph. This algorithm is a graph search one that solves the single-source shortest path problem for a graph with nonnegative edge path costs, producing a shortest path tree. It is used in routing and as a subroutine in other graph algorithms. It can also be used for finding the costs of shortest paths from a single vertex to a single destination vertex by stopping the algorithm once the shortest path to the destination vertex has been determined. For example, if the vertices of the graph represent cities and edge path costs represent driving distances between pairs of cities connected by a direct road, then Dijkstra's Algorithm can be used to find the shortest route between one city and all other cities. It can also be used by hospitals for ambulances in case of emergency to find the shortest available path. The algorithm creates a tree of shortest paths from the starting vertex, the source, to all other points in the graph. The algorithm exists in many variants; Dijkstra's original variant found the shortest path between two nodes, but a more common variant fixes a single node as the "source" node and finds the shortest paths from the source to all other nodes in the graph, producing a shortest-path tree [33]. For this, the following are points, which are necessary for Dijkstra's Algorithm in hospitals: [34]

- Existence of a widespread roads system that connects all parts of the city.
- Availability of sufficient VANET modules in the routes in order to detect traffic congestion.
- Infrastructure, such as GPS, communication links and two way radio are provided.
- Presence of a Dispatch Centre (DC) that serves the purpose of information exchange.
- Existence of an updated database of the roads and hospitals.
- Existence of Road Side Units (RSU) at suitable locations that might be inaccessible due to restrictions for propagation of signal.

E. Interface Design

In the proposed mHealth application, there are three potential users: rescuer or paramedic front-end, EDSO, and EMS administrator for the back-end. The rescuer or paramedic is the main user of the mHealth application. They can view a list of suitable nearest hospitals then choose one of them; they can also send the trauma-victim's assessment sheet containing key information on patients to the chosen hospital. On the other hand, the EDSO can view the assessment sheet of patients before they arrive (See Figure 3). An EMS dashboard administrator can view reports about every rescue/ paramedics operations and we can also search by rescue/ paramedics name, code number or hospital name (See Figure 4).

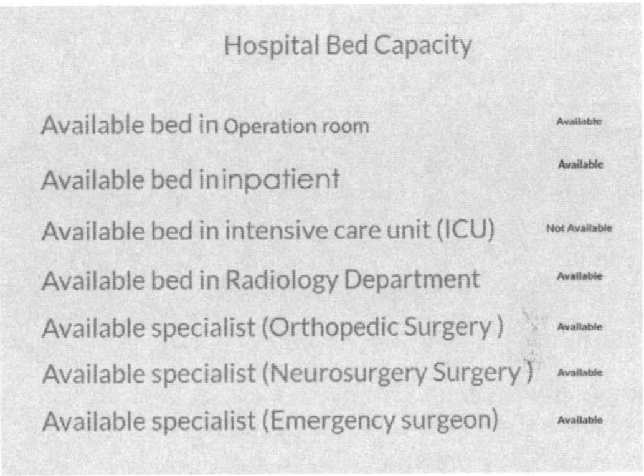

Figure 3. Hospital Bed Capacity Webpage Interface.

Figure 4. Emergency Dashboard Interface.

A menu in Figure 4 below shows all suitable hospitals arranged by location and available beds. This is a color-coded scheme of availability: Red icon: Not Available, and Green icon: Available.

- First icon from the right is: *Available bed.*
- Second icon from right is: *Available Orthopedic Surgery specialist.*
- Third icon from right is: *Available Neuro Surgery specialist.*

- Fourth icon from right is: *Available Emergency Surgery specialist.*

The screen in Figure 5 appears after choosing suitable hospitals from the previous screen. A *"field set"* of patient's attributes should be completed. A *Next* button transfers the user to the next page (confirmed page). A *Back* button transfers the user to the previous screen.

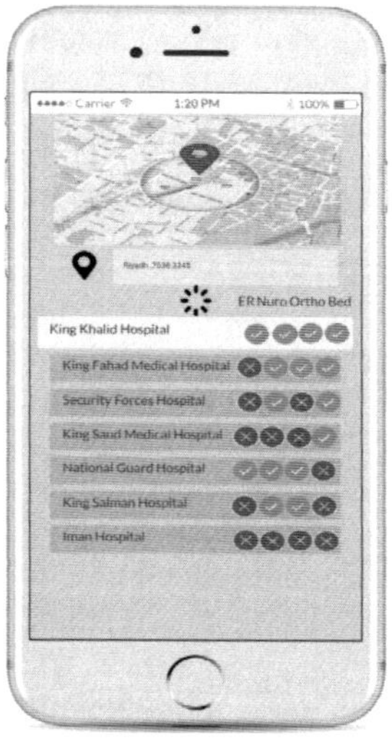

Figure 4. mHealth Application Interface (1st Screen).

Figure 5. mHealth Application Interface (2nd Screen).

V. Discussion and Conclusion

The work presented in this paper described the proposed solution to a major problem facing healthcare services. The proposed mobile health (mHealth) application solution is an attempt to solve the issue of major causes of deaths in trauma cases: ambulance diversion due to the unavailability of resources (namely beds and specialists) in the nearest hospital. The proposed application focuses on linking Emergency Medical Services (EMS) to trauma centers in Riyadh to help the paramedic to deliver the patients to the nearest trauma center with available resources in the shortest possible time to save the lives of patients. The evaluation results of the application have shown that the proposed solution provides regional-based availabilities of resources in nearest hospitals in order to avoid Emergency Departments (ED) crowding and shortages in inpatient bed capacity. The proposed application uses two algorithms: Dijkstra Algorithm for routing the ambulance path taken and Fusion Algorithm for predicting inpatient bed capacity. Results show an effective mHealth application to be used by 3 system users. First, EMS rescuer/ paramedic user to view a list of suitable nearest hospitals then chooses one of them; they also send the trauma-victim's assessment sheet containing key information on patients to the chosen hospital. Second, EDSO user to view the patients' assessment sheet before they arrive to the nearest trauma center that is ready to reduce the average time of 1 hour it takes for preparation prior to trauma victim arrival [31]. Finally, EMS administrator user to view a dashboard that reports about all EMS rescue/paramedics' actions. This mHealth application should optimize resources in EMS in Riyadh city to improve the ambulance diversion issue in particular and the quality of urgent care service delivery to Saudis in general.

Acknowledgment

The authors extend their appreciation to the Deanship of Scientific Research at King Saud University for funding this research through research group No (RGP-1438-002).

References

[1] Y. Shen, "Association Between Ambulance Diversion and Survival Among Patients With Acute Myocardial Infarction", *JAMA*, vol. 305, no. 23, p. 2440, 2011.

[2] "WannaCry ransomware: Everything you need to know", CNET, 2018. [Online]. Available: https://www.cnet.com/news/wannacry-wannacrypt-uiwix-ransomware-everything-you-need-to-know/. [Accessed: 11-Mar- 2018].

[3] Ministry of Foreign Affairs. "The Basic Law of Governance" Kingdom of Saudi. 1992.

[4] Ministry of Health, "Kingdom of Saudi Arabia - Ministry of Health Portal", 2017. [Online]. Available: http://www.moh.gov.sa/en/Ministry/Statistics/book/Pages/default.aspx. [Accessed: 16- Mar- 2017].

[5] Indexmundi. "Hospital bed density - Country Comparison", 2017. [Online]. Available: http://www.indexmundi.com/g/r.aspx?v=2227. [Accessed: 6-Mar- 2018].

[6] Van der Linden, M. C. "Emergency department crowding: Factors influencing flow",2015

[7] National Association of EMS Physici, "Ambulance Diversion and Emergency Department Offload Delay", Prehospital Emergency Care, vol. 15, no. 4, pp. 543-543, 2011.

[8] ACEP,"Definition of Boarded Patient",2017. [Online]. Available: https://www.acep.org/Clinical---Practice-Management/Definition-of-Boarded-Patient-2147469010/. [Accessed: 7- Mar- 2018].

[9] Encyclopedia. "bed occupancy - Dictionary definition of bed occupancy", Encyclopedia.com, 2017. [Online]. Available: http://www.encyclopedia.com/caregiving/dictionaries-thesauruses-pictures-and-press-releases/bed-occupancy. [Accessed: 7- Mar- 2018]

[10] B. Rechel, S. Wright, J. Barlow and M. McKee, "Hospital capacity planning: from measuring stocks to modelling flows," Bulletin of the World Health Organization, vol. 88, no. 8, pp. 632-636, 2010.

[11] World Health Organization, "Millennium Development Goals (MDGs), 2017. [Online]. Available: http://www.who.int/mediacentre/factsheets/fs290/en/. [Accessed: 7- Mar- 2018].

[12] M. McKee, "What are the lessons learnt by countries that have had dramatic reductions of their hospital bed capacity?" HEN synthesis report on reduction of hospital beds", Health Evidence Network (HEN), 2003.

[13] Jun, S. Jacobson and J. Swisher, "Application of Discrete-Event Simulation in Health Care Clinics: A Survey", The Journal of the Operational Research Society, vol. 50, no. 2, p. 109, 1999.

[14] Dawn M. Lewis," A Qualitative Case Study: Hospital Emergency Preparedness Coordinators' Perspectives Of

Preparing For And Responding To Incidents", Capella University, 2015

[15] R. Schmidt, S. Geisler and C. Spreckelsen, "Decision support for hospital bed management using adaptable individual length of stay estimations and shared resources", BMC Medical Informatics and Decision Making, vol. 13, no. 1, 2013.

[16] Chin-I Lin. "Optimization Models For Capacity Planning In Health Care Delivery." (Doctoral Dissertation). University of Florida, Gainesville, USA. 2008.

[17] B. Asplin, D. Magid, K. Rhodes, L. Solberg, N. Lurie and C. Camargo, "A conceptual model of emergency department crowding", Annals of Emergency Medicine, vol. 42, no. 2, pp. 173-180, 2003.

[18] Cisco Systems. "Singapore Hospital Modernizes Bed Management." [Accessed: 7- Mar- 2018] [Online]. Available: http://www.cisco.com/c/dam/en_us/solutions/industries/docs/healthcare/AlexandraHospital.pdf. 2006.

[19] Groundbreaking Bed Management Technology Helps SGH Improves Capacity and Care. [Accessed: 7- Mar- 2018] [Online]. Available:, from https://www.sgh.com.sg/about-us/newsroom/news-release/2010/Pages/GroundbreakingBedManagementTechnologyHelpsSGHImprovesCapacityandCare.aspx

[20] M. Ribbe, G. Ljunggren, K. Steel, E. Topinkova, C. Hawes, N. Ikegami, J. Henrard and P. JONnson, "Nursing Homes in 10 Nations: A Comparison Between Countries and Settings", *Age and Ageing*, vol. 26, no. 2, pp. 3-12, 1997

[21] Kerr E, Sibrand V. Health care systems in transition: Belgium. Copenhagen, European Observatory on Health Care Systems, 2000.

[22] S. Hussain, "Full Text of Saudi Arabia's Vision 2030 | Saudi Gazette", Saudi Gazette, 2017. [Online]. Available: http://saudigazette.com.sa/saudi-arabia/full-text-saudi-arabias-vision-2030/. [Accessed: 7- Mar- 2018].

[23] Ministry of Finance. "Ministry of Finance 2017 Budget of Kingdom of Saudi Arabia." [Online]. Available: https://www.mof.gov.sa/en/budget2017/Documents/The_National_Budget.pdf. [Accessed: 7- Mar- 2018].

[24] Ministry of Health. "Achievement of Ministry of health 2013", 2017. [Online]. Available: http://www.moh.gov.sa/en/ministry/statistics/book/documents/1433.pdf [Accessed: 7- Mar- 2018].

[25] Ministry of Health. "News - Ajyad Hospital in Makkah Shows Continuous Readiness for the Hajj Season", Moh.gov.sa, 2017. [Online]. Available: http://www.moh.gov.sa/en/hajj/news/pages/news-2014-09-22-002.aspx. [Accessed: 7- Mar- 2018].

[26] S. Walston, Y. Al-Harbi and B. Al-Omar, "The Changing Face of Healthcare in Saudi Arabia", Annals of Saudi Medicine, vol. 28, no. 4, pp. 243-250, 2008.

[27] A. Bagust, M. Place and J. Posnett, "Dynamics of bed use in accommodating emergency admissions: stochastic simulation model", BMJ, vol. 319, no. 7203, pp. 155-158, 1999.

[28] Emedicine.medscape. "Initial Evaluation of the Trauma Patient: Overview, Triage and Organization of Care, Initial Assessment", Emedicine.medscape.com, 2017. [Online]. Available: http://emedicine.medscape.com/article/434707-overview. [Accessed: 7- Mar- 2018].

[29] A. Bagust, M. Place and J. Posnett, "Dynamics of bed use in accommodating emergency admissions: stochastic simulation model", BMJ, vol. 319, no. 7203, pp. 155-158, 1999.

[30] Momammed Alouad, Business Analyst, Emergency Medical Service, Saudi Red Crescent Authority (SRCA), Riyadh, Saudi Arabia, 2017.

[31] Dr. Thamer Nouh, Trauma and Acute Care Surgery Unit, King Khalid University Hospital, King Saud University Medical City, Riyadh, Saudi Arabia, 2017.

[32] New South Wales (NSW) Health. "Emergency Department Support Officer", NSW Health eRecruit, 2017. [Online]. Available: http://nswhealth.erecruit.com.au/ViewPosition.aspx?Id=270427. [Accessed: 7- Mar- 2018].

[33] Yuemin Li, Renbiao Wu, Tao Zhang, "Multi-scale fusion and estimation for multi-resolution sensors", Signal Processing (ICSP) 2012 IEEE 11th International Conference on, vol. 1, pp. 193-197, 2012, ISSN 2164-523X.

[34] P. Singal and R. R.S.Chhillar, "Dijkstra Shortest Path Algorithm using Global Position System", International Journal of Computer Applications, vol. 101, no. 6, pp. 12-18, 2014.

An Enhanced Approach for the Prioritisation in Patent Ductus Arteriosus (PDA) Services Using Data Mining Techniques

Noura Alzeer, Hessah A. Alsalamah, Alaaeldin Hafez
College of Computer and Information Sciences, King Saud University
Riyadh, Saudi Arabia
e-mail: noura.alzeer@gmail.com, {halsalamah, hafez}@ksu.edu.sa

***Abstract*— The past few decades have witnessed a huge increase in the number of heart disease patients globally. An estimated 17.7 million people died from Cardio-Vascular Diseases (CVDs) in 2015, representing 31% of all global deaths. Saudi Arabia is no exception when it comes to the growing number of CVDs patients. The most common CVDs in Saudi Arabia is Patent Ductus Arteriosus (PDA), which is a persistent opening between two major blood vessels leading from the heart. Unlike many other fields of medicine, PDA is time-related and cannot be delayed. Thus, there is a need to have a Decision Support System (DSS) that can analyze the crucial statistics in deciding the level of urgency for each child that requires PDA surgery. This research proposes an approach to classify the PDA patients according to their urgency level based on their current conditions and the results of scans and tests they have undergone. An enhanced approach, which uses Data Mining (DM) techniques, is proposed to manage waiting lists and determine the priority among patients, as well as diagnosis of certain diseases. This approach aims to support early prevention and intervention to increase life expectancy and well-being of PDA patients. Our proposed approach achieved an accuracy of 99.15%.**

Keywords- Data Mining (DM); Healthcare; Patent Ductus Arteriosus (PDA); Waiting Lists Management.

I. INTRODUCTION

The alarming growth in pediatric and adult CVDs is now a major threat to world health [2]. The causes of CVDs include, but are not limited to diet through bad eating habits, obesity, and modern day lifestyle included stress. Other reasons may include general heart disease which can be hereditary, age and routine life pressure which can all be contributory to CVDs. There are at least eighteen distinct types of congenital heart defects that are recognized, with many additional anatomic variations [3].

This study focuses on one type, called PDA. In pre-term babies, the closure of the Ductus Arteriosus (DA) may be delayed after birth and is influenced by factors, such as gestation and development at birth, Respiratory Distress Syndrome (RDS), artificial mechanical ventilation, infection and lack of antenatal steroid treatment. These risk factors are more common amongst babies who require intensive care [2]. Persistent patency of the DA is therefore more common in pre-term infants; in over 50-60% of cases where babies are born at less than 29 weeks' gestation, the DA does not naturally close. This is known as a PDA [3].

In Saudi Arabia, physicians undertake a detailed medical examination of all pre-term infants, including examining the infant specifically for any PDA symptoms [4]. If any such symptoms are discovered, the newborn infant will receive immediate treatment. However, in full-term healthy babies, conditions, such as PDA are rarely checked. It is only if physicians or parents notice any PDA symptoms or if the child's non-diagnosed condition becomes critical to the point when it require prompt medical intervention that any treatment for PDA is given. Because of the increasing numbers of PDA patients in Saudi Arabia, it is important to raise the awareness of this disease and to educate parents, explaining the symptoms of this disease to ensure that patients are diagnosed as early as possible and thus receive the proper treatment as early as possible. It is also important for hospitals, once a patient is diagnosed with PDA, to ensure the patients have access to treatment in a timely manner. DM techniques, especially classification, have been used in many different medical fields, such as managing waiting lists and determining priorities among patients, as well as the diagnosis of certain diseases. In this research, we propose a hybrid approach of classification, clustering and association rules to determine not only life and death cases, but also cases which are both urgent and not urgent, and the urgency level within each category using the association rules. A DSS is used when scheduling patients or pre-planning a date for their scheduled treatment, procedure or surgery, and it is believed to have a positive impact on the current process.

The paper is organized as follows. Section II explains the importance of this work. Section III describes the theoretical background. Section IV presents related works to this research. Section V outlines the proposed research methodology, followed by Section VI, which shows how DM is used to build a model to diagnose PDA patients according to their urgency level. Section VII discusses the results obtained from the proposed model. Finally, the conclusion and future work of this research are found in Section VIII.

II. IMPORTANCE OF THE STUDY

Many guidelines have been published to determine the importance of diagnosing the urgency of PDA. This urgency can be summarized into two categories; life and death. Any delay in diagnosing the urgency could have a tragic fatal outcome, but early diagnosis and intervention can result in the increased life expectancy and well-being of the patient.

This study uses DM techniques to study and analyze PDA patients' records collected from King Fahd Medical City (KFMC) in Riyadh, Saudi Arabia, to build a new standard model to diagnose PDA and determine the urgency levels of PDA patients, depending on their existing medical condition, via results of scans and other tests carried out. It also presents recommendations for testing newborn babies for any PDA symptoms. This will help parents to receive the appropriate timely treatment for PDA for their infant before the condition worsens and requires surgery. The model is tested on new patients and its results are compared to those results provided by the physicians at KFMC to determine its accuracy. The model is also presented to further specialist heart physicians from different hospitals in Saudi Arabia to be tested. Their comments and recommendations are used to modify the original model to reach its final result. Having a DSS, which uses a standard model to provide an accurate urgency level for each PDA case, will save patients and their family time and money instead of trying to obtain various consultations from different physicians in different hospitals. Ultimately, there is also the cost-saving factor to the physician and hospital to be considered in reducing multiple appointments, as well as reducing multiple tests. The mere reduction in appointments alone opens up the availability of allocating an appointment to another patient. Overall, the hospital will also save large sums of money from the reduction of multiple examinations, tests and procedures, some of which are unnecessary as they may well be repetitive and thus not conducive to the well-being of the patient.

III. Related Work

Extensive research has been conducted studying the long waiting list problems, especially within medical fields. Some research focused on how to arrange priorities of these waiting lists and others dealt with how to reduce the general waiting times. Either way, what is definitive is that, generally, there is far greater demand over supply that results in waiting times and more extensive waiting lists.

A recent study [5] used statistical techniques to compare waiting lists for hip and knee arthroplasty at Groote Schuur Hospital in Cape Town in South Africa with the waiting lists kept by the surgeons at that hospital. The results showed that the hospital's waiting list is inaccurate because of the poor data management; for example, in many cases, the list did not have the current contact details of the patients. It also identified unfairness of waiting times among the patients. The research recommended the use of a scoring-based prioritization system which uses clinical diagnosis and tests results, radiographic and societal parameters to manage the waiting lists fairly.

A significant amount of research focused on using DM techniques in general in the healthcare sector [6]-[9] showed the benefits of applying different types of DM techniques in disease diagnosis and finding side effects of certain medicines, for example. However, they did not focus on one type of disease or on the type of DM techniques. Other studies used one or more DM techniques to predict or diagnose one type of disease; for example, recent research [10] used Naïve Bayes and WAC (Weighted Associated Classifier) to predict whether a patient has a risk of heart diseases or not. The study used factors, such as age, sex, whether diabetic, height, weight, blood pressure, cholesterol, fasting blood sugar and hypertension to predict the possibility of having heart diseases. In another study, the focus was on using DM techniques in oral cancer patients [11] and the authors showed that all the models built for predicting the survivability of oral cancer patients show similar results and performance. However, they did prove that the TreeBoost model is slightly better than the others as all 18 predictors they used in their research are considered for each spit. They showed the experimental results of probability adjustment, threshold analysis and lift-gain are also slightly better in the TreeBoost model. Therefore, they concluded that the TreeBoost classification model is an effective system for determining the survival rate of oral cancer patients. Rogers and Joyner [12] shows a special type of DM technique, called SAS software (Statistical Analysis Software), that can be utilized as a solution for the increasing costs in the healthcare sector and to improve the quality of service and care provided by hospitals and healthcare centers. The authors of the study discussed the DM methodology used in SAS software and applied it when studying the factors affecting the healthcare industry. It was shown that it can be used to solve serious and critical problems in the healthcare sector. This research is an example of applying types of DM (Statistical Analysis in this case) on medical data sets. Finally, in a closely-related study to this research topic [13], the authors submitted a survey of current techniques of knowledge discovery in databases, using DM techniques that are used in today's medical research, particularly in Heart Disease Prediction. When predicting a heart attack, 15 characteristics are listed and with basic DM technique, other approaches, such as Artificial Neural Network (ANN), Time Series, Clustering and Association Rules and soft competing approaches can also be combined.

This research focuses on using the basic information about the patient, such as age and results of scans, to provide an initial diagnosis of PDA. This helps to predict PDA cases as early as possible, especially in small towns which are far from large hospitals in major cities. It helps hospitals schedule urgent cases of PDA in waiting lists according to their urgency level.

IV. Theoretical Background

In this study, DM techniques are used to diagnose PDA and to arrange patients according to their urgency level. To gain a better understanding of this study, it is important to first understand what PDA is, what DM techniques are and how they work.

A. Patent Ductus Arteriosus (PDA)

PDA is "a persistent opening between two major blood vessels leading from the heart. The opening, called the ductus arteriosus, is a normal part of a baby's circulatory system before birth that usually closes shortly after birth. If it remains open, however, it is called a patent ductus arteriosus." [14]. This opening allows a great deal of blood to enter the lungs and heart, causing extra blood pressure on the lungs and enlarging and weakening the heart. There are no clear causes for this disease, but genetics may play a role. A defect in one or more gene could prevent the DA from closing normally after birth [14]. Symptoms of PDA depend on the size of the defect. While small PDA can have no signs sometimes, large PDA can be detected as a result of different causes. These include poor eating habits and consequently very slow growth, tiring and sweating during crying or eating, a rapid heart-beat and fast breathing [14]. Treatment for PDA can vary from medication or catheter-based procedures to surgery, depending on the age of the child and the size of the PDA [14] [15]. As PDA patient numbers increase, the pressure on hospitals to provide urgent care grows, whether this be interventional care or damage repair. Unlike other fields of medicine, in PDA, time is of essence and intervention cannot be delayed. Thus, the need to create a system that can speedily analyze the statistics is crucial in determining the level of urgency for each child that requires PDA-related surgery.

B. Data Mining (DM)

DM is a process, which finds useful patterns from large quantities of data [16]. It is the core step of the knowledge discovery process [16]. These steps are:

- Data selection: This is the selection of applicable data and records from the one or more databases applicable to the relevant study.
- Data preprocessing: This step involves the removal of irrelevant items and data entries that are not connected to the work environment. Data preprocessing allows the transforming of the original data into a suitable shape to be used by a mining algorithm, before starting the process of DM algorithm analysis [17].
- Data transformation and enrichment process: This step contains the calculating of the new characteristics from the existing characteristics.
- Data Mining: The application of DM techniques to the data to obtain new patterns and models.

DM can be carried out using numerous methods, depending on the type of data used and the result required from DM. In this study, the focus will be on three DM techniques: Classification, Clustering and Association Rule.

1) Classification:

As the main process [18] that is included in DM systems, this maps the various types of data into well-defined subgroups and classes. Another name for the classification process is Supervised Learning. Supervised Learning consists of two main stages; the first phase is a model construction and the second is model usage. Model construction: This phase contains the operations of building the used model in the structures of the attributes.

Model usage: All the operations that are used in utilizing the model.

2) Clustering:

Clustering is a process of grouping physical or abstract objects into classes of similar objects. Clustering and Classification are both grouping methods. Clustering is an unsupervised classification, whereas Classification is a supervised grouping. Classification and Prediction are also related techniques. Classification predicts class labels [19] whereas Prediction predicts continuous-valued functions. The major clustering algorithms are:

- Partitioning algorithms: Construct various partitions and then evaluate them according to certain criteria.
- Hierarchy algorithms: Create a hierarchical decomposition of the set of data (or objects) using certain criteria.
- Density-based: Based on connectivity and density functions.
- Grid-based: Based on a multiple-level granularity structure.
- Model-based: A model is hypothesized for each of the clusters and the idea is to find the best fit of that model for each other.

In this research, a portioning technique called K-Means is used [19]. This works with numerical values only, which suit the numerical measurements of the PDA scans.

3) Association Rule Mining (ARM):

The concept of ARM is considered to be one of the main studies in DM methods. Such rules associate one or more attributes of a dataset with another attribute, producing an if–then statement concerning attribute values. The original problem is the market basket analysis which tries to find all the interesting relations between the bought products. Sequential pattern mining attempts to find inter-session patterns, such as the presence of a set of items followed by another item in a time-ordered set of sessions or episodes. The association rule is an implication expression of the form X ➔ Y, where X and Y are disjoint item sets. The strength of the association rule can be measured by its support and confidence. Support determines how often a rule is applicable to a given dataset, while confidence determines how frequently items in Y appear in transactions that contain X [20].

V. Research Methodology

In this research, DM techniques are used to solve the problem of diagnosing and determining the urgency level of PDA patients. DM can assist in classifying patients according to the urgency of their cases in order to recommend the correct course of treatment, or pre-plan a date for their scheduled treatment, procedure or surgery. Within this study, data relating to PDA patients will be gathered from patients' records at KFMC, including their test results, image scans, medical history and physicians' diagnosis. This data will be stored in tables after being checked and cleansed to eliminate any incomplete records and to transform the data into numeric values or classify it into categories to prepare it for the application of different DM techniques.

Meetings with cardiology physicians at KFMC were initially conducted to discuss their approach to determine the urgency level for each PDA case. Physicians explained the main tests they carry out and scanned the images PDA patients have undertaken. Physicians provided a detailed explanation for each test and scan result, whilst showing the circumstances under which these results are considered normal, abnormal or critical. This helped gain an initial understanding of exactly how the urgency of PDA patient's urgency level is determined. Afterwards, DM techniques were applied to this data to find one standard model to classify patients according to their urgency level. In this research, Waikato Environment for Knowledge Analysis (WEKA) application was used to apply DM techniques. WEKA contains tools for data pre-processing, classification, regression, clustering, association rules, and visualization [21], which facilitates the application of different DM techniques on the data and compares the results in order to understand the relationships between the patients' attributes (details such as age, weight, height and test results) in order to obtain the most accurate model to classify patients according to the urgency level. Following this, testing was carried out on new patients using the proposed model. The results were compared to the results carried out by cardiology consultant physicians to ascertain its effectiveness and success. In case the proposed model yields inaccurate results, it was automated into a useful and easy to use tool. Otherwise, DM techniques were applied again to obtain an improved model, which was then tested again.

The proposed DSS can help in classifying the urgency level of PDA patients as speedily and efficiently as possible based on their test results, sometimes without the need to be examined, at this stage, by a physician. The measures to test the effectiveness of the system are based on two aspects:

- Time needed to determine the urgency level for a specific PDA case. Currently, physicians at KFMC meet for 30-60 minutes to consult on a single case and make a decision. Therefore, the new suggested tool should allow for considerably less time.
- Accuracy of the urgency level suggested by the tool. The tool will be used to determine the urgency level for new PDA patients. The consultant physicians at the hospital determine its accuracy and check the results given by the tool.

VI. Using DM to Build Model to Diagnose PDA Patients

In this research, DM techniques are applied on PDA patients' records.

A. Applying Classification on PDA Patient Records

After collecting PDA patients' records from KFMC, the data was stored in an Excel file in Comma Separated Values (CSV) format, which can be used in WEKA. The records included a copy of ultrasound scans photos. The details about the PDA dimensions are kept in a separate database, which is accessed, only by the physicians. The patient's record stores the PDA size as small, moderate or large. Attributes of the records are explained in Table 1.

After applying classification using decision tree classification, the tree in Fig. 1 was generated. The correctly classified instances are: 234, whilst the incorrectly classified are only 2 records, giving this model a very high accuracy percentage of 99.15%.

The main attribute for classifying the patients in the tree is the current age.

TABLE I. ATTRIBUTES IN PDA PATIENTS' RECORDS FROM KFMC

Attribute Name	**Type**	**Values**
Sex	Nominal	F or M
Current age	Numerical	Integers from 0 for new born
Height	Numerical	Integers
Weight	Numerical	Decimal numbers
Heart rate	Numerical	Integers
PDA last size	Nominal	**Non**: the patient does not have previous scan **Small, Mod, Large**
PDA current size	Nominal	**Small, Mod, Large**
Change	Nominal	**Yes** **No** **New**: if the patient does not have previous scan
Period to monitor	Numerical	Integers
Decision	Nominal	**Wait**: the patient gets medical treatment and waits for another scan after 3 or more months **Urgent**: if the patient needs Surgery or Catheter Procedure

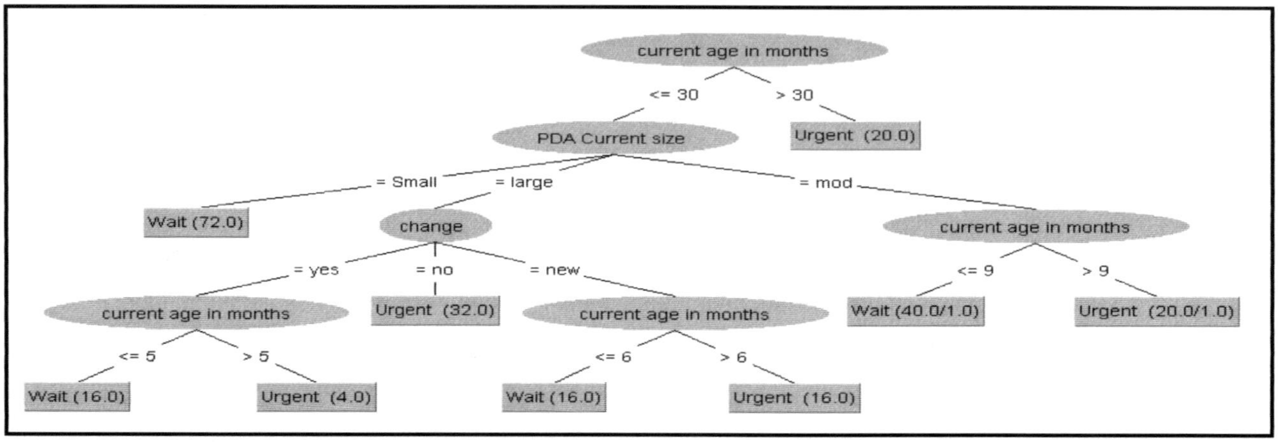

Figure 1. Decision Tree resulting from applying classification on PDA patients' records.

1. If the age of the patient is older than 30 months, then it is an Urgent case. This is applied in 20 cases.
2. If the child's age is less than, or equal to, 30 months and the current size of PDA is small then the patient receives medical treatment and waits for 3 months for a further scan. This is applied in 72 cases.
3. If the age of the patient is less than, or equal to, 30 months of age and the current size of PDA is moderate then there are two cases depending on the current age:
   - If the age is less than, or equal to, 9 months then the patient receives medical treatment and waits 3 months for another scan. This is applied in 40 cases, with the exception of one case where the decision was revealed as Urgent.
   - If the age is more than 9 months, then the patient case is urgent. This is applied in 20 cases, except for one case whereby the decision was to receive medical treatment and wait for 3 months for a further scan.
4. If the age is less than, or equal to, 30 months and the current size of PDA is large then there are 3 cases depending on whether there was a change on the PDA size.
   - No change: the case is classified as Urgent. This is applied in 32 cases
   - There was change: there are 2 cases based on the age:
     i. Age less than, or equal to, 5 months: receive medical treatment and wait 3 months for new scan. This is applied in 16 records
     ii. Age is greater than 5 months: it is an Urgent case. This is applied in 4 cases.
   - The change = New, which denotes a new patient with no previous record. There are 2 cases based on the current age:
     i. Age less than, or equal to, 6 months: receive medical treatment and wait 3 months for new scan. This is applied in 16 cases.
     ii. Age is greater than 6 months: it is an Urgent case. This is applied in 16 cases.

The previous model uses the PDA size and patient age to provide an accurate initial diagnosis of PDA.

B. Applying Clustering on PDA Patient Records

One of the clustering techniques provided by WEKA is called Simple K-Means [21]. WEKA uses the mean of the attributes to group the records into K groups or clusters, such that the records in each group are closer to each other (similar to each other) than they are to records in other groups. Clustering works better with numerical attributes than with nominal attributes. The records of PDA patients collected from KFMC have both numerical attributes, such as, age, height and weight, and nominal attributes, such as, PDA size, and gender. In order to apply a Simple K-Means algorithm, nominal attributes should be changed into numerical attributes.

The specialists from KFMC who were consulted for this study explained that gender is not a major attribute in deciding the urgency level of PDA patients. Research and publications of study articles related to PDA treatments also suggest that although PDA is more common in girls than boys, gender does not affect the diagnosis of the disease. For these reasons, sex attribute is not used in the clustering technique. The other nominal attribute is PDA size, and it is replaced with the average size as follows:

- Small less than 1.5mm. Therefore, it is replaced by 1.25.
- Moderate between 1.5mm and 2mm, so it is replaced by 1.75.
- Large greater than 2mm, so it is replaced by 2.25.

Applying simple K-Means making K = 2, which means the results will give 2 clusters, one for patients who receive treatment and wait for 3 months to have another scan and a class for patients who need urgent treatment. The two clusters are shown in Table 2.

TABLE II. RESULTS OF APPLYING CLUSTERING ON PDA PATIENTS' RECORDS

Attribute	Cluster #1	Cluster #2
Current age in months	15.5109	7.0625
Height	71.9565	62.7451
Weight	10.6685	6.9181
Heart rate	115	108.3333
PDA last size	1.2391	1.125
PDA current size	1.9457	1.6111
Decision	**Urgent**	**Wait**
Number of records	92	144

The table shows that:

- Older patients (average 15.5 months) are more likely to be classified as urgent, while younger patients (average 7.1 months) receive treatment and remain under observation.
- As expected, older patients have higher height and higher weight, so height and weight averages are related to the same clusters as age.
- Higher heart rate was related to Urgent cases while lower heart rate is related to patients who receive treatment to remain under observation.
- The important attribute here is the current size of PDA: average 1.9 (which means moderate or large PDA) means Urgent case, while smaller PDA with average 1.6 is not urgent.

These results comply with the results obtained from the decision tree used in the classification.

C. Applying Association Rule on PDA patient records

Association rule technique is applied to those patients whose cases are classified as "Urgent" to determine the urgency level. Doctors determine the urgency level of the patient based on many factors, such as age, size of PDA, tests and ultrasound scans. There are three levels of urgency:

- Level1: needs immediate surgery.
- Level2: needs surgery within a month or two.
- Level3: needs to undergo more tests and scans to determine a time for surgery.

To apply the association rule, all the attributes should be nominal, so the patient's age is transformed into age group. Also, the heart rate was transformed from numbers into nominal: low, normal, and high. The height and weight were also transformed into nominal values.

- Group a: less than 6 months
- Group b: between 6-12 months
- Group c: between 12-36 months
- Group d: older than 36 months

These ranges are based on the physicians' suggestion at KFMC. After applying association rule on the patients' records, the following rules were obtained:

- PDA Current size =large 103
  ==> urgency level=level1 103
- Age group=a 95
  ==> urgency level=level1 95
- Age group=a PDA Current size =large 39
  ==> urgency level=level1 39
- Age group=b PDA Current size =large 30
  ==> urgency level=level1 30
- Age group=b PDA Current size =mod 12
  ==> urgency level=level2 12
- Age group=c PDA Current size =mod 51
  ==> urgency level=level1 51
- Age group=c PDA Current size =large 45
  ==> urgency level=level1 45
- Age group=c PDA Current size =small 64
  ==> urgency level=level3 64
- Age group=d 72
  ==> urgency level=level1 72
- Age group=a change =no 39
  ==> PDA Current size =large 39

All the above rules have a confidence = 1, which means they are 100% true based on the PDA patients' records from KFMC. Table 3 shows a summary of these rules.

TABLE III. RESULTS OF APPLYING ASSOCIATION RULE DM ON THE URGENT CASES OF PDA PATIENTS

Urgency level	Cases
Level1	PDA Size = Large or Age is less than 6 months or Age between 12-36 months and PDA size = Mod or Age older than 36 months
Level2	Age between 6-12 months and PDA size = Mod
Level3	Age between 12-36 months and PDA = small

VII. RESULTS

DM classification and clustering techniques were used to diagnose PDA and classify patients into two categories: urgent cases and non-urgent, which are referred to here by Wait cases where the patient receives medical treatment and waits for another check-up after 3 or more months. Urgent cases were then arranged into three levels of urgency using association rule DM techniques. The models presented by this research were tested by cardiology specialists at KFMC who subsequently approved them. They suggested using age range instead of actual age in months to make the diagnosis more general. The suggested age groups are:

a: 0-6 months / b: 6-12 months / c: 12-36 month / d: older than 36 months

After applying the specialists' recommendations, the resulting model is shown in Table 4.

TABLE IV. THE RESULT OF APPLYING DM TECHNIQUES ON PDA PATIENTS' RECORDS

Case Description	Diagnosis	Urgency level
Age group = a, b & PDA current size = Small	Wait	
Age group = a & PDA current size = Large & Change = yes	Wait	
Age group = b, c & PDA current size = Large & Change = yes	Urgent	Level1
Age group = a, b, c & PDA current size = Large & Change = no	Urgent	Level1
Age group = a & PDA current size = Large & Change = new	Wait	
Age group = b, c & PDA current size = Large & Change = new	Urgent	Level1
Age group = a & PDA current size = Mod	Wait	
Age group = b & PDA current size = Mod	Urgent	Level2
Age group = c & PDA current size = Mod	Urgent	Level1
Age group = c & PDA current size = Small	Urgent	Level3
Age group = d	Urgent	Level1

VIII. CONCLUSION AND FUTURE WORK

This study presented a hybrid approach to diagnose PDA as early as possible using simple tools. The introduction of an ultrasound scan that is readily available in almost all medical centers in Saudi Arabia is one method of detection and diagnosis. This study also recommends increasing the awareness of PDA amongst first time parents, as well as parents with more than one child, as early as possible, to ensure that they are educated to understand what PDA is and to seek medical help as soon as they notice any symptoms in their baby, such as shortness of breath, sweating while feeding, struggling to breathe and nurse at the same time or quite simply being too attached to nursing. To a parent, some of these symptoms may not appear to be any obvious cause for alarm, but with prior knowledge of PDA, early diagnosis and intervention could quite well positively impact on health and life expectancy. Ultimately, the goal is to produce a longer living and healthier Saudi population.

It is expected that when increasing the awareness of PDA, this will, in turn, increase the number of PDA patients in Saudi Arabia. The result will create a strain on already struggling hospitals and also substantially increase the number of patients needing to go onto a waiting list to see a consultant as well as those patients who have already received a diagnosis but require ongoing medical treatment to prevent surgery, as well as to maintain their health. Overall, the burgeoning hospital outpatient clinics for those patients with PDA will need to be overhauled to ensure that all patients are attended to in timely and appropriate manner.

While working on this research, we faced many difficulties. For example, the details of the scans, such as the dimensions of the PDA, were not saved in the patients' records. This information was saved in special files which can be accessed by the physicians only. This made collecting the information needed for the patients extremely difficult. It was not easy to meet the specialists at KFMC for any length of time as they were tremendously busy all the time.

Regarding future work, further study and analysis should be carried out on more patients from different hospitals to obtain a more accurate model to diagnose PDA and determine its urgency level. Additional data should be added to the patients' records to show all the detailed measurements taken from the ultrasound scan.

ACKNOWLEDGMENT

The authors extend their deep appreciation to the Deanship of Scientific Research at King Saud University for funding this work.

REFERENCES

[1] World Health Organization. [Online]: http://www.who.int, [February 2018].

[2] World Heart Federation. [Online]: http://www.world-heart-federation.org [February 2018].

[3] National Heart, Lung, and Blood Institute, "What is the heart?" [Online]: http://www.nhlbi.nih.gov/health/health-topics/topics/hhw/, [February 2018].

[4] Saudi Ministry Of Health website, "Cardio-Vascular Diseases". [Online]: http://www.moh.gov.sa/HealthAwareness/EducationalContent/Diseases/Heartcirculatory/, [February 2018].

[5] T. Kavalieratos, M. Nortje, and R. N. Dunn. "Hip and knee arthroplasty waiting list - how accurate and fair?" *South African Medical Journal*, Apr 2017.

[6] R. Zhang and Y. Katta. "Medical Data Mining, Data Mining and Knowledge Discovery," in *SPIE*, 2002, pp. 305-308.

[7] H. Wang and S. Wang. "Medical Knowledge Acquisition through Data Mining," in *IEEE*, 2008, pp. 1-4.

[8] L. Passmore, J. Goodside, L. Hamel, L. Gonzales, T. Silberstein, and J. Trimarchi, Assessing decision tree models for clinical in-vitro fertilization data. Technical Report TR03-296, Dept. of Computer Science and Statistics, University of Rhode Island, 2003.

[9] S. Patel and H. Patel, "Survey Of Data Mining Techniques Used In Healthcare Domain." *International Journal of Information Sciences and Techniques*, vol. 6, 2016.

[10] A. Patel, S. Gandhi, S. Shetty, and B. Tekwani. "Heart Disease Prediction Using Data Mining." *International Research Journal of Engineering and Technology*, vol. 4, 2017.

[11] N. Sharma and H. Om, "Data mining models for predicting oral cancer survivability," *Netw. Model. Anal. Heal. Informatics Bioinforma.*, vol. 2, no. 4, pp. 285-295, 2013.

[12] G. Rogers and E. Joyner, "Mining your Data for Health Care Quality Improvement," *SAS Institute*, Inc., January 2009.

[13] J. Soni, U. Ansari, and D. Sharma, "Predictive Data Mining for Medical Diagnosis: An Overview of Heart Disease Prediction." *International Journal of Computer Applications,* March 2011.
[14] Mayo Clinic, [Online]: http://www.mayoclinic.org/diseases-conditions/patent-ductus-arteriosus/basics/definition. [February 2018].
[15] University of California, Department Of Surgery. [Online]: http://pediatric.surgery.ucsf.edu/conditions--procedures/patent-ductus-arteriosus.aspx, [February 2018].
[16] Mrs. Bharati M. Ramageri, "Data Mining Techniques and Applications", *Indian Journal of Computer Science and Engineering,* vol. 1, no. 4, pp. 301-305.
[17] W. Klosgen and J. Zytkow, *Handbook of data mining a knowledge discovery.* Oxford University Press: New York, 2002.
[18] B.K. Bharadwaj and S. Pal, "Data Mining: A prediction for performance Improvement using classification", *International Journal of Computer Science and Information Security*, vol. 9, pp. 136-140, 2011.
[19] J. Han, M. Kamber, and J. Pei, *Data Mining: Concepts and Techniques*, Morgan Kaufmann Publishers, San Francisco, 3rd edition, 2013.
[20] R. Agrawal, T. Imielinski, and A. Swami, "Mining association rules between sets of items in large databases," In Proceedings of the 1993 ACM SIGMOD international conference on management of data, pp. 207–216, 1993.
[21] WEKA. [Online]: www.weka.com, [February 2018].

RoboChain: A Secure Data-Sharing Framework for Human-Robot Interaction

Eduardo Castelló Ferrer*[1], Ognjen (Oggi) Rudovic*[1], Thomas Hardjono[2], Alex 'Sandy' Pentland[1,2]
[1]MIT Media Lab, Massachusetts Institute of Technology, Cambridge, MA 02139, USA
[2]MIT Connection Science & Engineering, Massachusetts Institute of Technology, Cambridge, MA 02139, USA
Email: {ecstll, orudovic, hardjono, pentland}@mit.edu
*these authors contributed equally to this work

Abstract—**Robots have potential to revolutionize the way we interact with the world around us. One of their largest potentials is in the domain of mobile health where they can be used to facilitate clinical interventions. However, to accomplish this, robots need to have access to our private data in order to learn from these data and improve their interaction capabilities. Furthermore, to enhance this learning process, the knowledge sharing among multiple robot units is the natural step forward. However, to date, there is no well-established framework which allows for such data sharing while preserving the privacy of the users (e.g., the hospital patients). To this end, we introduce RoboChain - the first learning framework for secure, decentralized and computationally efficient data and model sharing among multiple robot units installed at multiple sites (e.g., hospitals). RoboChain builds upon and combines the latest advances in open data access and blockchain technologies, as well as machine learning. We illustrate this framework using the example of a clinical intervention conducted in a private network of hospitals. Specifically, we lay down the system architecture that allows multiple robot units, conducting the interventions at different hospitals, to perform efficient learning without compromising the data privacy.**

Keywords–Distributed Robotics; Data Privacy; Blockchain; Federated Learning; Distributed Robotics; Mobile Health Technologies.

I. INTRODUCTION

Recent advances in mobile and robotic technology have found applications in many domains including entertainment, education and health [1]. In particular, Socially Assistive Robotics (SAR) [2] has emerged as a research field that aims to create robots that can empower humans in a number of activities. This has brought various types of social robots: from humanoid robots to medical devices and responsive home appliances [3]. One of the main potentials of this type of robots is their ability to monitor and improve human well-being and health [4]. More specifically, in the health domain, robots have been used to improve clinical interventions for individuals with neurodevelopmental conditions such as autism [5], and also for monitoring and assisting people with conditions such as dementia [6], among others. As part of the intervention, robots need to be able to establish naturalistic and engaging interactions with humans. Since these robots are typically equipped with a number of sensors including audio-visual sensors, and also have access to vast 'prior' data (different types of interactions tested in different contexts), they have potential to constantly learn and improve their interaction capabilities [7]. This, in turn, could lead to more effective health and well-being interventions as the robots will constantly be improving by learning from data. This is also true in the case of alternative mobile health technologies [8] (e.g., mobile phones, tablets and computer monitors), however, the three dimensional embodiment of robots is typically perceived more human-like and engaging by target patients [9].

Until recently, this process was limited due to the inefficiency of existing learning techniques (e.g., most of human-robot interactions would be pre-scripted). The increase in available human data ('big data') and progress in Machine Learning (ML) [10] (in particular, 'deep learning' [11]), have enabled to automate parts of health interventions [12], allowing the robots to learn more efficiently and customize the interventions. However, this requires the use of personal and highly sensitive data, especially when working in clinical settings [13][14]. The main downside of existing solutions is that these data are highly isolated, thus, not shared among different sites (e.g., hospitals). This is mainly because of data privacy issues and their potential misuse by untrusted parties. This limits the knowledge sharing that could otherwise benefit clinical interventions enhanced by robots. Consequently, this constraints the learning and adaptation of the robots to existing and new contexts, hindering the progress toward more efficient health interventions. Even though the SAR is moving rapidly, to date, there is no a well-established framework that addresses the challenges mentioned above, while exploiting the full potential of personal and other sensitive data in an open, secure, decentralized, and efficient manner.

Over the last decade, it has been shown that open access datasets have been extremely useful to extract knowledge in diverse fields such as urban planning [15], environmental monitoring [16], and health care [17]. However, as the ubiquity of these data increases, it might also become less secure. In contrast, siloing data can provide a measure of security, but raises issues of inter-operability [18]. There is also increasing interest by governments [19], societies [20], and industry [21] to share knowledge based on data. However, this needs to be balanced with the right to preserve the privacy of the subjects represented in the data [22]. To provide a viable solution to this trade-off, novel platforms like MIT OPen ALgorithms (OPAL) [23] have proposed a change of paradigm: rather than moving data into a centralized location, so that it can be queried, analyzed, and processed by an algorithm, the queries are delivered to the nodes containing the datasets of interest instead. In other words, the algorithm needs to be 'sent' to the data. Then, queries would be executed by the relevant node, with the results being reported back to the querier – who would merge the results into a meaningful analysis. In this way, the raw data never leaves its physical location and the owner never looses control over it. Instead, nodes that carry relevant datasets execute queries and report the results. Importantly, security and privacy becomes more manageable in this paradigm because each node controls its own data store and monitors the privacy entropy of released answers.

Likewise, the emergent technologies such as blockchain [24], a chronological ledger of transactions that ensures the integrity of the information included, can be used to capture and log both queries and its correspondent answers. Blockchain provides a useful mechanism to support post-event audits [25], enhanced privacy, and availability [26] in the systems that rely on sensitive data. These characteristics of the system increase its transparency and have already been proven useful in medical applications [27][28], as well as distributed robotics scenarios [29][30]. Moreover, the blockchain technology offers practical means to safely and securely store and track the use of personal data as well as the parameters of the ML models used for the robot intervention. This increases the users' trust in the system and provides a rich source of information that can be used to better design future interventions. It is also worth noting that our framework deploys blockchain technology but is independent of any specific implementation. We seek to cater to a broad set of deployment scenarios. As such, RoboChain can be deployed with (i) public blockchains (ii) semi-private blockchains, or (iii) private/permissioned blockchains. We define a public blockchain as one where anyone can read/write to the blockchain, and as such has the ability to read and validate transaction entries. We define a semi-private blockchain as one where anyone can read and validate transaction entries, but only authorized entities are able to create or write transaction to the blockchain. Finally, we define a private/permissioned blockchain as one where only authorized entities are able to read/write to the blockchain. In the following, we simply refer to the blockchain regardless of the type used.

To address the challenges of the private data inclusion in Human-Robot Interaction (HRI) and, more specifically, as part of health/clinical interventions, we propose a secure and efficient framework aimed at addressing the following aspects of target HRI interactions: (i) how to achieve an effective and efficient data-driven HRI based on patients' data from clinical interventions without breaching the data privacy. This includes mechanisms to notarize, verify, and account for all inflows/outflows of the data involved in the process. (ii) How to achieve an efficient way to train and continuously improve ML models, being a part of the robot's perception during the intervention [13], using the interaction data collected in (i). (iii) Finally, we address how to efficiently update and share the learned ML models obtained in (ii) among different robot units connected in a sparse network (e.g., multiple hospitals). To this end, we introduce a novel framework for secure data-driven HRI, named RoboChain, which builds upon and combines the latest advances in OPAL, blockchain, and ML technologies. We illustrate its potential utility in the context of health domain and clinical interventions as part of autism therapy. However, the framework is applicable to any HRI where the use of personal data and their sharing is critical for the task. The key to the RoboChain approach is that users (e.g., patients) have the possibility to check what information was generated and/or captured during the interaction with the robot (e.g., the therapy), and confirm that this information do not compromise their privacy. Furthermore, we show how the secure sharing of the knowledge of the robots connected in a network can increase the efficacy of a decentralized learning process.

The rest of the paper is structured as follows. Section II presents an overview of how assistive robots carry out health interventions and the different elements involved in the process. Section III describes in detail the RoboChain architecture and its information workflow. Section IV describes the combination of RoboChain and federated learning. Finally, we conclude this work in Section V with the discussion, limitations and future work.

II. Assistive Robots for Health Interventions

To illustrate our approach, we use as a running example the HRI in an occupational therapy ("the intervention") for individuals with autism ("the patient"), where a humanoid robot NAO ("the robot") is used as an assistive tool. Engaging patients with autism in an intervention is a challenging problem as they easily lose their interest and often quickly disengage from the intervention activities. However, the majority of these patients find robots quite engaging because of their, in contrast to humans, consistent behavioural expressions. This, in turn, allows them to sustain the patient's engagement and perform the intervention more effectively. The latter is achieved via a number of pre-scripted activities by the robot, aimed at improving socio-cognitive skills of the patients. Namely, these patients have, among others, challenges in interpreting behavioural cues of emotions of other individuals. The goal of the robot intervention is then to assist the therapist in teaching the patients to recognize/imitate these expressions. For instance, as part of the intervention, the robot may play an imitation game with a patient by asking the patient to show his/her expression of joy. Then, the robot conveys the same emotion via his voice and bodily movements programmed to show the expression of joy of typical individuals. In this way, the patient learns to identify typical expressions of various emotions, with the aim to use that knowledge in future interactions with his/her peers. In what follows, we outline three key elements of the robot-assisted intervention that need be considered in the proposed RoboChain architecture.

A. Interaction Model (IM)

This refers to the set of activities and behaviours, defined by the domain experts, that need be performed by the robot as part of a target intervention for autism. Specifically, IMs rely on the high-level interaction modules controlling the execution of target behaviours by the robot. These behaviours are implemented to simulate the typical steps that an experienced therapists would perform as part of an intervention. However, an IM also contains the robot sensing and perception modules, which deploy pre-trained ML models to automatically estimate the key metrics needed to modulate the intervention. These may include the models for automatic detection of the patient's low-level behavioural cues such as head pose and gaze direction, as well as high-level metrics such as engagement levels. To this end, the robot uses locally stored and pre-trained ML models. For instance, the robot can use Deep Learning Models (DLM) [11] designed for detection and interpretation of the patient's behavioural cues directly from the image data recorded using a robot-embedded camera. However, the key challenges in this process are: (i) how to efficiently update the IMs, and thus their DLMs, (ii) how to safely share these models across multiple sites (e.g., hospitals) and (iii) how to assure that the current IMs/DLMs are the best among existing ones for analysis of the patient's target behavioural cues.

B. Interaction Data (ID)

This refers to the data that can be harnessed using the robot's embedded sensors such as cameras and microphones, as well as from other sensors installed at the site. As these encode personal data of the interacting patient, they should not leave the site in their raw form as it can compromise the patient's privacy. On the other hand, ID pose a great value for improving the performance of the ML models (DLMs) used by the robot. Furthermore, by consolidating the data from multiple sites, more effective IMs/DLMs can be built from these large datasets, compared to the data accessible only locally (e.g., within a single hospital). Briefly, one way to improve the models without sharing the raw data across different sites is to re-train the IM/DLMs at the target site, and then share the models' updates with the other sites. To improve the models, experienced therapists/clinicians provide feedback, for instance, on how well the robot performed during the therapy, also sometimes in terms of manual correction of the robot's estimates of the metrics of interest for the therapy (e.g., engagement levels for a specific patient). Once this information is accumulated locally in the form of the Therapist Feedback (TF) data paired with the raw ID, the IMs/DLMs can be updated/re-trained. For instance, this can be accomplished by fine-tuning the DLM weights using local processing servers or even on the robot hardware. The next key step is to securely share this newly acquired knowledge across multiple sites.

C. Background Data (BD)

In contrast to the ID acquired as the therapy progresses, the BD concerns the medical records of the patient, along with his/her demographics, and any other relevant (contextual) information. The latter may be the result of health and behavioral assessments of that particular patient, including, for instance, the family history, school reports, previously used medication and diagnoses, etc. These are typically stored as part of government and other public institution data, including educational centres and schools. However, instead of providing patient-identifiable data to the robot, these sites rather provide aggregated data, thus preserving the data privacy. By having access to these data and the local expert knowledge, the robot can customize its IM/DLMs to the target patient. In what follows, we describe how each of these key components (IM/DLMs, ID and BD) can efficiently and safely be used within the RoboChain framework to optimize the intervention, and, thus, its outcomes and the patient's experience.

III. The System Architecture

Figure 1 lays out the communication architecture of an individual robot deployed within a clinical intervention on the target site. This architecture outlines how the robot can gain access to safe BD about the target patient in order to adapt the therapy. It also shows how the access to this information is accounted for, and finally, how the resultant IM can be shared among different robot units in the RoboChain to increase their usability. We break down the proposed data flow into three main components as described below.

A. Background Data Service

① - The robot is equipped with a list of secure 'vetted' algorithms to build its queries. The goal here is to ensure that the algorithms used are free from any kind of bias (e.g., the culture and gender discrimination, or other types that do not comply with ethical norms). They should also prevent any unintended side effects (e.g., the exposure of the patient's behavioral or condition severity). To accomplish this, the 'vetted' algorithms must be verified beforehand by domain experts. For instance, the robot may query the general physical condition of the patients undergoing similar clinical interventions in order to adapt the interactions, which otherwise could adversely affect the patients. It is also important to note that this vetting does not guarantee the quality of the output which is a function of the quality of the input data. As an example, a set of hospitals that the robot is intending to query about the patient may possess detailed information about the mental wellbeing of similar patients, but may not return the BD in its original form. Rather, it provides the information about what to be avoided during the intervention instead of providing the exact details to the robot.

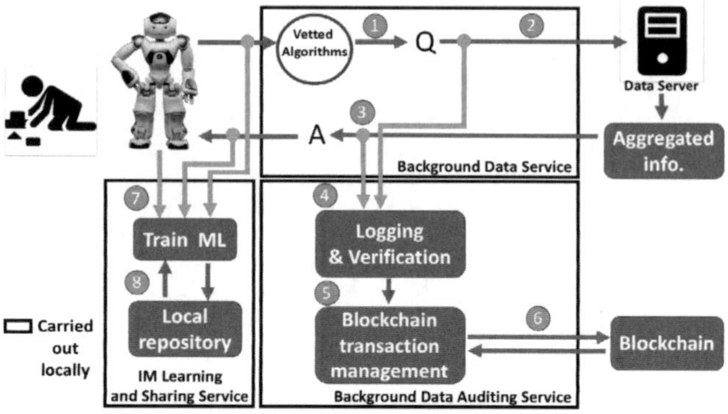

Figure 1. System architecture and data flow. The proposed system involves three main sections where (A) safe data is retrieved from protected databases using the OPAL paradigm, (B) the pairs of queries/answers are stored in the blockchain for accountability and transparency reasons, (C) the data derived from the therapy is used as an input to train and share ML models. All the procedures within the black rectangles are carried out in a local fashion (i.e., at the robot's own hardware).

② - Queries using one of the 'vetted' algorithms are sent to the data server. These queries may contain information obtained during the therapy. Specifically, the queries may be designed based on the expert input (by the caregiver or therapist) and/or by the robot's sensing of the environment including the patient. For the former, the caregiver administering the intervention at the site may inform the robot about the patient's condition and the type of information needed to execute/choose the current intervention. For our running example, i.e., the autism therapy, the therapist can decide that the patient needs to work on specific skills (e.g., practicing the eye-gaze exchange). The robot will then try to retrieve relevant information (through the queries of BD) that can assist him to better adapt the target IM to this type of social exercise.

To date, the majority of patients' BD are stored in different datasets scattered across the Internet: belonging to governments, public institutions, private corporations, etc. Consequently, the most recent advances in network data analysis are starting to address the challenge of creating value from those datasets without breaching anyone's privacy. OPAL is a framework that proposes a change of paradigm: instead of copying or sharing raw personal data, algorithms in the form of queries are sent to the datasets containing the personal information/data. Then, the queries are executed behind existing firewalls and only anonymous/aggregated information is sent back to the querier. In Figure 2, a complete patient profile which may include the patient's medical, educational, and/or governmental data, is employed to assist the intervention. Each one of these databases is controlled by a trusted party, and personal and sensitive data is stored in a 'raw' format from which the BD is retrieved. Then, queries are sent to each target database and aggregated information is returned (Figure 2). As mentioned in Section II-C, the aggregated information contains the group-level information about the patient (e.g., demographics such as culture, age and gender, and aggregated behavioral assessment scores). This information allows the querier (i.e., a robot) to obtain knowledge about the target patient needed to select the most appropriate IM/DLMs. The key here is that from the returned information, the patient cannot be identified nor his/her personal data compromised during the knowledge exchange. In addition, there is a number of parameters that the robot can passively observe prior to forming the query. For instance, the robot can use computer vision algorithms to automatically infer the patient's age, gender, and the motor abilities.

Figure 2. Data flow: Trusted parties (e.g., doctors, educators, public institutions, etc.) have access to the Background Data (BD) of patients. The BD are stored in protected databases that accept queries from trusted data servers. Data servers receive the 'answers' to those queries in the form of aggregated information. Aggregation is a useful mechanism to anonymize the information returned.

③ - An answer with aggregated information is received at the robot's side. Then, the robot can tailor the IM based on this. For instance, suppose that the answer is given in the form: "For this type of patients, the propensity to negative emotional reactions is $n\%$ higher than in other groups when exercise x is applied." By knowing this, the robot can adjust the IM and, therefore, minimize the risk of exposing the patient to experiencing negative emotions.

B. Background Data Auditing Service

④ Before the question-answer pair, which is obtained during the background data service phase, can be passed to the robot, the background auditing service section of the proposed model captures the pair to archive it on the blockchain.

⑤ Once the question-answer is stored, the system checks how to include this information in a blockchain transaction. This requires to encapsulate the information within the right data structure, depending on the type of blockchain employed. Note that here (a) the blockchain only holds the hash-value, and (b) the complete question-answer is kept by the data service where the raw data is located (as most of today's blockchains cannot encrypt large amounts of data).

⑥ The pair of queries-answers used in the system is included in the blockchain. Then, anyone auditing the clinical interventions can inspect the pair of questions-answers used during the patient's therapy. This increases the accountability and transparency of the HRI system, thus, improving its trustworthiness with all parties involved.

C. IM/DLM Learning and Sharing Service

⑦ As explained in Section II-A, the robot relies on the DLMs, being part of the target IM, when conducting a clinical intervention. Specifically, the robot checkouts the current DLM from its local hub (Figure 4 (A)) and uses it to conduct the intervention. During this, the robot stores the information captured by its sensors, producing the ID (e.g., the audio-visual recordings of the patient and his/her responses to the IM). Together with the available BD and TF, these data allow the robot to update/improve the existing DLM models. It is worth to note that updating the IMs is also feasible at this point, however, it requires more input from the domain experts as it concerns the intervention protocols. Specifically, a supervised ML approach is adopted: the ID/BD are used as input to the DLMs, while the TF (e.g., patient's engagement levels) as the target output. Then, the fine-tunning of the DLM parameters to the newly acquired data is accomplished using the standard back-propagation technique and by selecting an optimizer (e.g., Adadelta) [11]. With these new parameters, the robot is expected to increase its competences and adaptability to target interventions and patients [13]. An example of the data flow to/from IM and its DLMs is depicted in Figure 3.

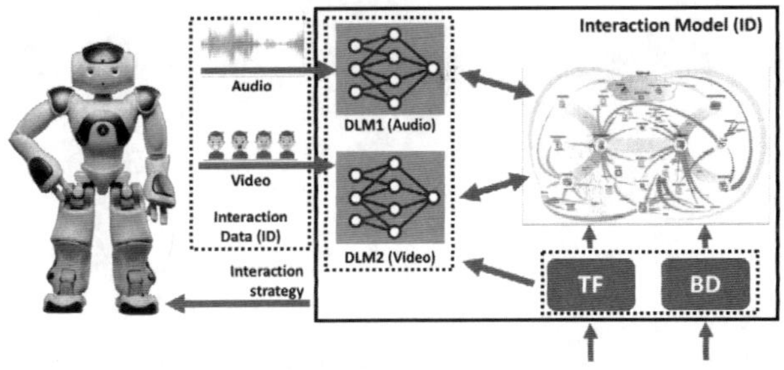

Figure 3. Interaction Model (IM). In this example, the robot provides audio-visual recordings of the patient as input to the IM. These are then processed using target DLMs, being part of the existing IM. The outputs of the DLMs (e.g., the patient's stress level and other behavioral cues) are fed into the IM scripts designed for the target intervention. This is further enhanced by the BD and, later, TF, both of which are then used to re-train (update) the DLMs. The output of the IM is the optimal interaction strategy to be performed by the robot as part of an ongoing intervention.

⑧ The patient data (ID, BD, and TF) collected by robot units at a single site (e.g., a hospital) are being stored at a local hub (Figure 4). In this scenario, a hub represents the computing infrastructure of the hospital, care center, etc., where the robot is performing the intervention. After the intervention, the patient data and TF are used together to improve the IMs by re-training their DLMs (as described in ⑦). This can be done directly on the hardware of local robot units, and/or on the data-processing servers of the local hub. To this end, different learning strategies can be adopted. For instance, the robots can create, store, and update the target models after each intervention, or after a sufficient amount of data has been collected. The design of specific learning strategies depends on the intervention type, and it is out of the scope of this paper. Then, these new models are committed and stored in the local repository (Figure 4 (B)) of the local hub. Once created and stored, it is assumed that these new IM/DLMs cannot be used to recreate the raw input data (ID) of the patients. Therefore, all personal data (e.g., images, audio, etc.) remain safe as they do not leave the hub. Furthermore, they are deleted after the models are updated.

The stored IM/DLMs are further deployed locally and assigned a cumulative score based on the TFs derived by validation of this model within new local interventions. As a result, a new Candidate Model (CM) along with its performance score is then created and locked on the local hub. To allow knowledge sharing – one of the key ingredients of the RoboChain framework – this new CM is then evaluated by the robot peers operating at other sites, i.e., different hubs within a clinical network. To this end, the local repository (hub) publishes the changes (Figure 4 (C)) (e.g., the difference between the previous version of the model and the new one). Finally, the hub announces the update to the entire network (Figure 4 (D)). The goal of this is to assure a fair validation of the CM before it can be adopted as the new version of the IM/DLMs for target intervention.

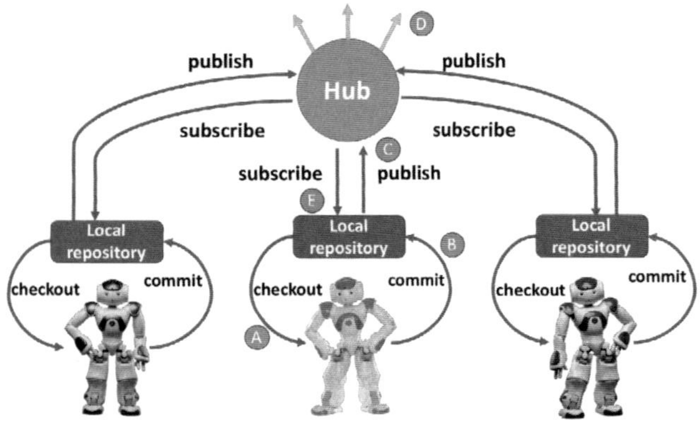

Figure 4. A group of robots commit/checkout changes from/to their local repository. Robots publish/subscribe to local hubs in order to send/get new updates on their DLM. One of the main advantages of this approach is that learning always take place locally, while the resultant knowledge is distributed globally. Having a local repository with the capabilities of a modern source control system (e.g., git or mercury) allows robots to calculate differential changes on the DLM structure, topology, etc., but also being able to roll back to previous versions.

IV. RoboChain and Federated Learning

The key property of the RoboChain framework is its ability to perform ML operations of the IMs/ DLMs without the need to store the data, acquired during interventions, in a centralized location (e.g., a shared hub within a network of hospitals). This is achieved using the notion of "Federated Learning" (FL) [31], which allows for smarter models, lower latency, and less power consumption, while ensuring the patients' privacy. Furthermore, FL allows the new models to be deployed immediately on the robot units from other sites such as different hospitals in a private network.

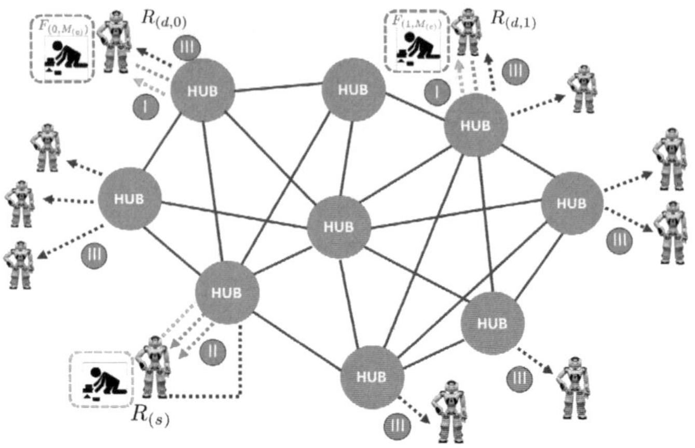

Figure 5. A global RoboChain network. Robots connect to local hubs, which are interconnected in a sparse network. Robots publish and receive notifications whenever new therapy sessions are conducted. CMs are proposed by individual robots and validated by the network in future therapies. After feedback is provided by different peers, a consensus process starts and the utility of the CM is evaluated. If the CM outperforms previous solutions, the model is acknowledged as the new standard.

Figure 5 depicts the FL approach employed in the RoboChain network. First, as mentioned in Section III-C, the source hub/robot $(R_{(s)})$ 'advertises' the new CM by announcing the IM/DLM updates to the entire network. Then, the destination robots $(R_{(d,i)})$, where $i = 1 \ldots N$ denotes the target sites/hubs, are notified by their local hub that there is an update available in the network (Figure 5 (I)). This can be achieved by the subscription pipeline they have with their local hub (Figure 4 (E)). In case the updates are available, the robots can retrieve and apply them to their working directory.It is worth to note that this does not mean that the robots are committing the received updates to their local repository yet. Once the CM $(M^s_{(c)})$ is adopted from the source hub by the destination robots, the latter will start its evaluation, quantified by the therapists at the destination sites in terms of the TF scores $(F_{(R_{(d,i)},M^s_{(c)})})$. Additionally, in order to leverage the new local data, the destination hubs can also return the model updates to the source site (obtained in a similar fashion as when creating the CM). These, in turn can be used to construct the new model at the source hub. An example of this approach, but applied in the context of mobile phones, can be found in [31].

The next stage in RoboChain is to consolidate the feedback information from the destination hubs/robots (Figure 5 (II)). This can be achieved using time-constraints (i.e., waiting for a pre-defined period of time to receive the feedback), and/or when a target consensus is achieved. For instance, if the feedback score for the CM $(M_{(c)})$ is higher than for the currently accepted model $(M_{(j)})$, i.e., $\overline{F_{(i,M^s_{(c)})}} > \overline{F_{(i,M_{(j)})}}$. If this is fulfilled, $R_{(s)}$ creates a new model $(M_{(j+1)})$, which is then published to all connected hubs, and committed to their robots' local repositories (Figure 5 (III)). In this way, the new baseline model for future interventions is endorsed by the network. This process can be implemented via modern control version systems (e.g., git, mercurial, etc.) in order to store and share the resulting model configurations (e.g., the DLM topology, hyper-parameters, etc.) obtained after new interventions. This is an important feature of RoboChain since it allows the robots to rollback to the last consensual version of the model ($M_{(j)}$), in case a consensus did not take place within the network. Moreover, since the robots keep the list of all changes in their local repository, there is a promising research approach in analyzing the metadata available in the updates applied to the repository.

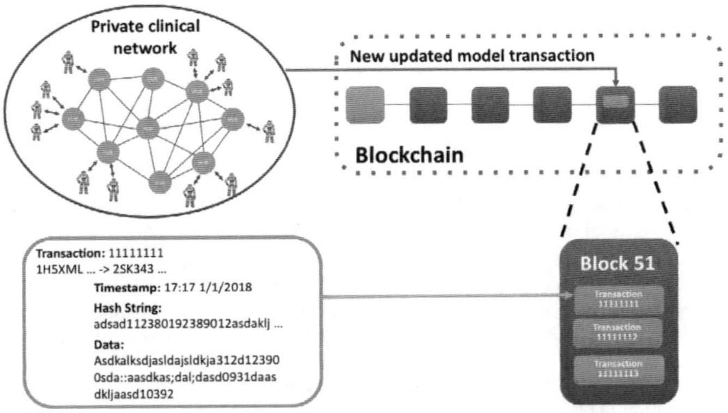

Figure 6. Once a $R_{(s)}$ obtains a new candidate model validated by the network, $R_{(s)}$ is responsible to send a transaction to a blockchain including the key information to verify the consensus process. Important information such as timestamps, hash of the new model, and encrypted data is provided. Even though this transaction is included in a public blockchain, the information within is only readable by the participants of the private clinical network.

In order to notarize and log the creation of the new models and their consensus processes within the network (Figure 6), $R_{(s)}$ is required to send a transaction to a blockchain (public, semi-public, or private) including information such as the timestamp of the global update broadcast (Figure 5 (III)), a hash string that encapsulates the information about the model update (e.g., differences in the weights, hyper-parameters, etc.), and an encrypted data field signed by $R_{(s)}$ containing information such as the public IDs of the robots that took part in the consensus process and their correspondent feedback scores. This transaction on the blockchain is necessary to allow participants within the private network (e.g., hospitals, care-centers, etc.) to prove/validate how models were created, who participated in their consensus process, and when did those transaction take place. It is also important to highlight that the hash string included in this transaction is useful to check and confirm that the models acquired by robots are indeed the version agreed upon by the network, and not a corrupted version from a third-party agent. In addition, the encrypted data field signed by $R_{(s)}$ contain sufficient information to allow the users to confirm their participation in the consensus process and check that $R_{(s)}$ was not biased at the moment of promoting $M_{(j)}$. This information is readable by any participant in the private clinical network, since they know the public identifier (i.e., public key) of $R_{(s)}$. By contrast, the peers connected to the blockchain but not members of the private clinical network do not have the means to decrypt this information or identify the nodes involved in the consensus process. This is because all required public/private keys remain within the boundaries of the clinical private network. Finally, note that the whole consensus reaching process could have been implemented directly on the blockchain via 'smart contracts' in order to prevent intruders from 'attacking' the network; yet, we assume here that the network access is protected.

V. Discussion, Limitations and Future Work

Until recently, the creation and training of DLMs have been a computationally-hungry process (e.g., requiring expensive dedicated hardware such as GPU chips). Because of this, the adoption of these models in embedded devices (e.g., robots) has been a challenging task, especially for low-cost robot units. However, not long ago, new platforms such as TensorFlow Lite [32] have paved the way for deployment of machine learning inference algorithms directly on target devices. This has been made feasible via an improved scheduling process. For instance, the DLM training happens only when the device is idle, plugged in, and counts with an open connection to the Internet, so there is no impact on the robot's main tasks. Tools such as TensorFlow Lite provide a lightweight solution for DLMs allowing also the federated learning approach, as proposed in this paper.

The federated learning network interconnecting the participant parties (e.g., hospitals, care centers, etc.) and its correspondent robot units is not public. Thus, it requires a permission of one (e.g., governmental institutions) or several parties to join and start contributing to the DLM learning process. We opted for this design approach since we understand that some form of trust is required in the institutions that deal with a sensitive task such as autism therapy. For these reasons, we assume there are no byzantine peers (e.g., malicious partners) within the network trying to bias or hinder the system operation. However, due to the inclusion of both transactions in the blockchains, the OPAL query/answer pair and the announcement of a new DLM adoption, we provide the necessary means to allow all the participants to check and prove the integrity of the interaction and learning process in RoboChain. Note also that our current framework assumes that the new models (IM/DLMs) are published by one party at the time; yet, multiple sites can propose new candidate models at the same time. This could easily be managed by introducing 'smart contracts' on the blockchain to attain the synchrony among the sites. Also, we illustrated our framework using a single intervention, but the RoboChain should leverage data from multiple interventions occurring simultaneously at different sites. To tackle this, IMs/DLMs could be structured in a modular form, enhancing each other as more patient data become available.

One of the aspects left for future work is the adaptation of the proposed system to different ethical frameworks for working with human data [33]. As mentioned in Section III-A, the use of 'vetted' algorithms ensures that the queries created during the therapy are fully aligned with the ethical policies of the relevant regulatory bodies (e.g., Institutional Review Boards (IRB)). This is particularly important when working with healthcare data [34], as envisioned in RoboChain. It is also important to emphasize that in RoboChain both queries and answers are stored on a blockchain instance, which allows all interested parties to verify the compliance of these ethical norms. However, further research is needed to develop robot controllers that protect the safety ("by default") of the patients (e.g., due to hardware malfunctioning and/or software bugs). For instance, robot controllers should always ensure that regardless of what type of information/DLMs are obtained through the RoboChain, the physical and mental states of the patients should not adversely be affected in any way.

To summarize, in this paper, we proposed RoboChain - the first learning framework for tackling the privacy issues in the use of personal data by multiple robots during HRI. We illustrated this framework using autism therapy as an example of an intervention conducted simultaneously by robot units at multiple hospitals. While the RoboChain proposed here offers the main principles for secure data and models sharing between multiple robot units and their sites, in future we plan to empirically evaluate this concept. To this end, in the first stage we aim to collect the intervention data from several hospitals in order to test the RoboChain framework in a simulated data-exchange scenario. In the second stage, we aim to have the RoboChain system running in real time on a private network. One of the main challenges that we envision in this process is how to enable efficient and real-time learning of IMs/DLMs. We expect that for this, existing ML approaches will need to be adapted so that they can efficiently communicate/be integrated with the OPAL and blockchain technologies. Also, how to extend the RoboChain framework so that it can simultaneously handle multiple private networks, in order to further increase the learning efficacy or interface with mobile health devices, are other promising directions to pursue.

ACKNOWLEDGMENT

This project has received funding from the European Union's Horizon 2020 research and innovation programme under the Marie Skłodowska-Curie grant agreement No. 751615 and No. 701236.

REFERENCES

[1] H. I. Christensen, A. Okamura, M. Mataric, V. Kumar, G. Hager, and H. Choset, "Next Generation Robotics," Tech. Rep., jun 2016.

[2] M. J. Matarić, "Socially assistive robotics: Human augmentation versus automation," Science Robotics, vol. 2, no. 4, mar 2017, p. eaam5410.

[3] B. Adams, C. Breazeal, R. Brooks, and B. Scassellati, "Humanoid robots: a new kind of tool," IEEE Intelligent Systems, vol. 15, no. 4, jul 2000, pp. 25–31.

[4] L. D. Riek, "Healthcare robotics," Communications of the ACM, vol. 60, no. 11, oct 2017, pp. 68–78.

[5] O. Rudovic, J. Lee, L. Mascarell-Maricic, B. W. Schuller, and R. W. Picard, "Measuring Engagement in Robot-Assisted Autism Therapy: A Cross-Cultural Study," Frontiers in Robotics and AI, vol. 4, no. 36, jul 2017.

[6] M. Valentí Soler et al., "Social robots in advanced dementia," Frontiers in Aging Neuroscience, vol. 7, no. 133, 2015.

[7] P. Liu, D. F. Glas, T. Kanda, and H. Ishiguro, "Data-Driven HRI : Learning Social Behaviors by Example From Human – Human Interaction," IEEE Trans on Robot., vol. 32, no. 4, 2016, pp. 988–1008.

[8] C. Free et al., "The Effectiveness of Mobile-Health Technologies to Improve Health Care Service Delivery Processes: A Systematic Review and Meta-Analysis," PLoS Medicine, vol. 10, no. 1, 2013, p. e1001363.

[9] W. A. Bainbridge, J. W. Hart, E. S. Kim, and B. Scassellati, "The benefits of interactions with physically present robots over video-displayed agents," International Journal of Social Robotics, vol. 3, no. 1, 2011, pp. 41–52.

[10] M. I. Jordan and T. M. Mitchell, "Machine Learning: Trends, perspectives and prospects," Science (New York, N.Y.), vol. 349, no. 6245, 2015, pp. 255–260.

[11] Y. Lecun, Y. Bengio, and G. Hinton, "Deep learning," Nature, vol. 521, no. 7553, 2015, pp. 436–444.

[12] OECD, Data-Driven Innovation Big Data for Growth and Well-Being: Big Data for Growth and Well-Being. OECD Publishing, 2015.

[13] O. Rudovic, J. Lee, M. Dai, B. Schuller, and R. W. Picard, "Personalized machine learning for robot perception of affect and engagement in autism therapy," arXiv:1802.01186, 2018.

[14] S. Alsalamah, W. A. Gray, J. Hilton, and H. Alsalamah, "Information security requirements in patient-centred healthcare support systems," Studies in Health Technology and Informatics, vol. 192, no. 1-2, 2013, pp. 812–816.

[15] J. Dong, W. Cheng, and L. Xin, "Data Augmented Design: Urban Planning and Design in the New Data Environment," in 2017 IEEE 2nd International Conference on Big Data Analysis (ICBDA), 2017, pp. 508–512.

[16] B. L. Sullivan et al., "Using open access observational data for conservation action: A case study for birds," Biological Conservation, vol. 208, no. May, 2017, pp. 5–14.

[17] P. Kostkova et al., "Who Owns the Data? Open Data for Healthcare," Frontiers in Public Health, vol. 4, no. February, 2016.

[18] R. M. Frey, T. Hardjono, C. Smith, K. Erhardt, and A. S. Pentland, "Secure sharing of geospatial wildlife data," in Proceedings of the Fourth International ACM Workshop on Managing and Mining Enriched Geo-Spatial Data - GeoRich '17. New York, New York, USA: ACM Press, 2017, pp. 5:1–5:6.

[19] J.-N. Lee, J. Ham, and B. Choi, "Effect of government data openness on a knowledge-based economy," Proceedings of Computer Science, vol. 91, no. Supplement C, 2016, pp. 158–167, promoting Business Analytics and Quantitative Management of Technology: 4th International Conference on Information Technology and Quantitative Management (ITQM).

[20] B. Wessels, R. Finn, K. Wadhwa, and T. Sveinsdottir, Open Data and the Knowledge Society. Amsterdam University Press, Mar 2017.

[21] G. Li, Developing the Data Knowledge Industry. Springer Berlin Heidelberg, 2011, pp. 84–96.

[22] J. C. Alsalamah, Shada, Gray, William Alexander and Hilton, "Sharing patient medical information among healthcare team members while sustaining information security," in 15th International Symposium for Health Information Management Research (ISHIMR 2011). Zurich: University of Zurich, 2011, pp. 553–554.

[23] T. Hardjono, D. Shrier, and A. Pentland, Trust::Data: A New Framework for Identity and Data sharing. Boston: Visionary Future, 2016.

[24] S. Nakamoto, "Bitcoin: A peer-to-peer electronic cash system," [retrieved: March, 2018]. [Online]. Available: http://bitcoin.org/bitcoin.pdf

[25] S. Suzuki and J. Murai, "Blockchain as an Audit-Able Communication Channel," in Proceedings - International Computer Software and Applications Conference, vol. 2, 2017, pp. 516–522.

[26] X. Liang, S. Shetty, D. Tosh, C. Kamhoua, K. Kwiat, and L. Njilla, "ProvChain: A Blockchain-Based Data Provenance Architecture in Cloud Environment with Enhanced Privacy and Availability," in Proceedings - 2017 17th IEEE/ACM International Symposium on Cluster, Cloud and Grid Computing, CCGRID 2017, 2017, pp. 468–477.

[27] A. Azaria, A. Ekblaw, T. Vieira, and A. Lippman, "MedRec: Using blockchain for medical data access and permission management," in Proceedings - 2016 2nd International Conference on Open and Big Data, OBD 2016, 2016, pp. 25–30.

[28] Q. Xia, E. B. Sifah, K. O. Asamoah, J. Gao, X. Du, and M. Guizani, "MeDShare: Trust-Less Medical Data Sharing Among Cloud Service Providers via Blockchain," IEEE Access, vol. 5, 2017, pp. 14 757–14 767.

[29] E. C. Ferrer, "The blockchain: a new framework for robotic swarm systems," arXiv:1608.00695, 2016.

[30] V. Strobel, E. Castello Ferrer, and M. Dorigo, "Managing Byzantine Robots via Blockchain Technology in a Swarm Robotics Collective Decision Making Scenario," IRIDIA – Technical Report Series, Brussels, Tech. Rep., 2018.

[31] H. B. McMahan, E. Moore, D. Ramage, S. Hampson, and B. A. Arcas, "Communication-Efficient Learning of Deep Networks from Decentralized Data," arXiv:1602.05629, 2016.

[32] "Tensorflow lite," [retrieved: March, 2018]. [Online]. Available: http://www.tensorflow.org/mobile/tflite/

[33] J. Metcalf and K. Crawford, "Where are human subjects in Big Data research? The emerging ethics divide," Big Data & Society, vol. 3, no. 1, jan 2016, pp. 1–14.

[34] B. C. Stahl and M. Coeckelbergh, "Ethics of healthcare robotics: Towards responsible research and innovation," Robotics and Autonomous Systems, vol. 86, 2016, pp. 152–161.

So'rah: An Arabic Mobile Health Application for Saudi Dietary Evaluation

Abeer S. Al-Humaimeedy, Rahaf Almozaini, Lama Almansour, Kholoud Alaqeely, Ahlam Almutairi, Abeer Alolayan

Department of Software Engineering
King Saud University
Riyadh, Saudi Arabia
Emails: Abeer@ksu.edu.sa, Rahaf.almozaini@gmail.com, lmoo-2@hotmail.com, {Khoulod.abdullah,Ahlamalmutairi744,Abeer9471}@gmail.com

***Abstract*—Current studies show that obesity is reaching higher rate in Saudi Arabia. Health problems related to obesity, like diabetes, obstructive sleep apnea, and osteoarthritis are becoming serious concerns in the Saudi society. In the same time, mobile phones have been embraced by a rapidly increasing number of people worldwide in general and in Saudi Arabia in particular. Using smartphone mobile health applications to watch food calories is an effective method to fight obesity. However, most of the current mobile diet applications are available in English language and do not provide the number of calories for Saudi foods. This paper presents *So'rah*, a mobile health application for a dietary evaluation. *So'rah* aims to show the number of calories in Saudi foods based on barcode reading using a mobile camera, or typing full or part of a food's name. Moreover, the application allows guest (unregistered) users to explore the reviews for different kinds of foods. If users register in the application then they can review the taste and quality of the food and create a diet plan to maintain their weight by logging intake calories and follow up drinking water. In this paper, we will describe the development process of *So'rah* mobile application and the implementation challenges.**

Keywords–Mobile health; mHealth; Application; Dietary evaluation; Obesity.

I. INTRODUCTION

Saudis through time are becoming more concerned about their health and their eating lifestyle. This concern comes in the fact of Saudi Arabia now has one of the highest obesity and overweight prevalence rates [1]. Studies show that 7 out of 10 Saudis are overweight [2]. This is a direct result of Saudi Arabia becoming more developed. Populations in the developed world are mostly affected by obesity. The fast and busy style of modern life prevent people from the pleasure of thinking of the quality of their food and choosing the healthy ingredients [2].

Obesity is a major source of a number of diseases, including hypertension, diabetes, obstructive sleep apnea, hyperlipidemia, and osteoarthritis [1]. Previous studies have revealed that obesity is also among the major cause of co-morbidities, including cardiovascular diseases, cancers, and the related issues that may lead to morbidity and mortality [2].

In addition to the health problems, obesity represents a relative economic burden on countries economy [3].

Obesity is considered as a significant public health issue, which raise a concern globally. The World Health Organization (WHO) states that obesity rates have been nearly tripled worldwide, since 1975 [4]. Statics conducted in 2016, by WHO organization shows that 39% of the adults were overweight and 13% were obese [4]. The same statics show that 41 million children under age 5, around the world, were either overweight or obese [4].

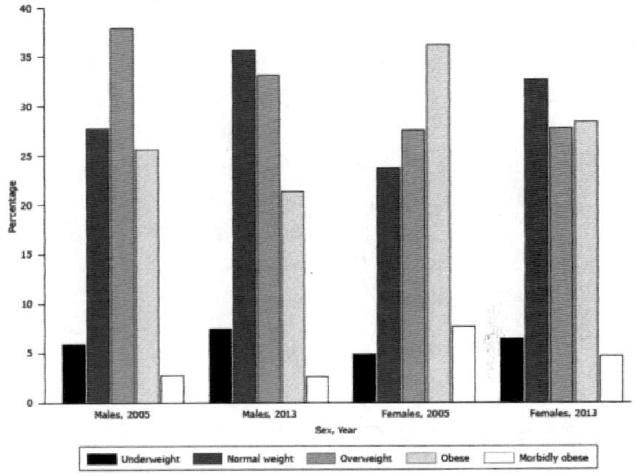

Figure 1. Overweight percentage changes from 2005 through 2013, for men and women in Saudi Arabia. [5]

Nowadays, Saudis become more aware that healthy eating habits will lead to a healthy life. The conducted survey in [5] shows that, although the overweight prevalence rates are high in Saudi Arabia, the overweight prevalence rates are decreasing, as shown in Figure 1.

Moreover, mobile phones have been embraced by a rapidly increasing number of people worldwide, in general, and in Saudi Arabia, in particular.

Therefore, using smartphone mobile health applications can contribute to decrease the overweight prevalence rates. B. Spring *et al.* [6] shows that, people who use mobile weight loss applications as a part of a comprehensive weight loss program, are better able to lose weight and keep it off.

There are several applications that have been developed to help people looking after their health. One type of these applications is calories counter applications. Users of these applications can keep track of the intake calories in their meals by calculating the number of calories in a specific amount of the consumed food. Some applications provide extra features like follow up users exercises and drinking water.

However, most of the applications in the mobile markets are in English and not including Saudi foods in their official lists. Consequently, Saudi users for these applications will not be able to keep track of the number of calories in their everyday meals because their consumed food are not there.

Therefore, in the paper, we present a mobile health application, namely "*So'rah*", which is designed for Saudi users. The application aims to help the community to lower the rates of obesity and overweight in Saudi Arabia by providing the Saudi users with the information they need about calories in Saudi foods either by searching food lists using the name of the consumed food or scanning foods' barcode. To the best of our knowledge our application is the first Saudi mobile application which identifies foods by scanning their barcodes. In addition, the application supports extra features like reviewing foods in the application lists, creating diet plans for users to help them maintain their weight, and enhancing the application food lists by allowing users to add new foods. Food additions will only appear in food lists if they have been approved by the application nutrition supervisor; for details see Section III-A.

Paper overview: Section II discusses related work. Section III introduces *So'rah* mobile application. Section IV demonstrates and design the main features of *So'rah* mobile application. Finally, Section V concludes and discusses the application future directions.

II. Related Work

In this section, we first review some of the trending calories counting applications according the heath websites [7] and [8] and a number of online magazines like [9] and [10]. Secondly, we compare them to the proposed features of our application.

Lose It [11], *MyFitnessPal* [12], *FatSecret* [13], and *Weight Watchers* [14] applications are well known weight loss applications which keep track of calories consumption for users, and help them to set their ideal weight as a goal and monitor their progress towards it. The food databases of these applications can be searched using a food's name. *Lose It*, *MyFitnessPal*, and *FatSecret* applications allow users to search food lists by scanning food's barcode. Moreover, *Lose It* and *FatSecret* applications can recognize eaten foods by their pictures.

The main feature of *Weight Watchers* application is that the recommended daily meals are under the supervision of certified nutritionists and dietitians, which makes the application like online diet clinic [14]. However, the application is not free and require a monthly subscription.

Food databases (lists) of these applications are comprehensive, e.g., *MyFitnessPal* has more than 5 millions foods in its database [12]. However, none of the previously mentioned applications have Saudi foods in their databases, and in case they have, like *MyFitnessPal*, then it is marked as a user addition which is not validated and approved by the application administrators.

To overcome this limitation some endeavours have emerged in the Arabic mobile stores.

Soraate [15], *Calorie counter from my diet* [16], *Calories guide* [17], and *My diet* [18] applications are examples of these endeavours. They are Arabic applications that help users to maintain their weight by following up the gained calories from everyday meals and the burned calories in daily exercises. Foods and drinks can be selected from applications lists of Arab foods which are updated regularly. Water and follow up drinking feature are also supported in some of these applications.

My diet application is considered as online diet clinic like *Weight Watchers* application. Users can consult certified nutritionists and dietitians to construct a customized diet plan. However, *My diet* application is also not totally free.

Food databases (lists) of these applications include Arab foods, which are not necessary to be Saudi foods, e.g., searching for *Jarish*, a traditional Saudi food, can only be found in *Soraate* application. Additionally, foods in these applications' databases are identified by name only. Therefore, scanning barcodes to identify foods are not supported.

Our application *So'rah* includes Saudi foods in its database and identifies foods by name and by barcode. Therefore, searching for Saudi foods using keywords or barcodes both are supported.

Table I shows a brief comparison between our *So'rah* application and the previously listed applications. The selected features for comparison are chosen based on the main functionalities of these applications.

It can be concluded from Table I that the implemented features of *So'rah* application cover most of the listed features except following up exercises, and synchronizing with health applications and/or smart watches to automatically get the users gained/burned calories. We consider these missing features important, but we preferred in the current version of *So'rah* to concentrate in developing a mobile application with a comprehensive database which will facilitate its use in the Saudi society. The missing features are implemented in the second version of *So'rah* which is currently under development.

Additionally, it can be noticed from Table I, that our application allows users to add new foods and in the same time it is considered to be full supervised by professionals. This comes as a result of prohibiting a food addition from appearing in food lists until it is approved by the application nutrition supervisor. On contrary, food additions in the other applications, require no approval, which could generate a miss leading information if the new addition is incorrect.

Final point, users of *So'rah* application can write reviews about foods in our lists, and like/dislike previously written reviews. These reviews can be comments on the food itself, like its taste, or can be a space for fellow users to exchange recipes, eaten experiences or recommend restaurants for this particular food. Therefore, the reviews section can be considered as a reviews section and a chatting forum in the same time. However, with the reviews section, users can locate and track fellow users chats about a particular food because they know that all reviews about a food appear in its information page.

A detailed discussions and justifications of *So'rah* features will be presented in the next section.

III. Application Construction

This section discusses the application main features and briefly discusses its design. It also presents a system architecture for our application which foster its practical deployment.

TABLE I. SUMMARY OF THE APPLICATIONS.

Feature	*So'rah*	*Cal. counter*	*Soraate*	*mDiet*	*Cal. guide*	*Lose It*	*FatSecret*	*MyFitnessPal*	*Weight Watcher*
Calories for Saudi foods	√	partially	√	partially	partially			not validated	
Search by food name/barcode	√	name	name	name	name	√	√	√	√
Log exercises		√	√	√		√	√	√	
Sync. with health apps						√	√	√	√
Sync. with Smart watches						√		√	
Water tracking	√	√	√	√		√		√	
Support Arabic language	√	√	√	√	√				
Food reviews and rating	√		rating						
Chatting forums		√	√	√		√	√	√	√
User can add new food	√		√					√	
Application is free	√	√	√	partially	√	partially	√	partially	
Supervised by professionals	√	√	partially	and consultancy	√	√	√	partially	and consultancy

A. *The Application main Features*

In *So'rah* application, we have four types of users as following:

- **Guest user**: A user who does not register in the application but wants to use some of the application features.
- **Registered user**: A user who does register in the application and can use the full features of it.
- **Nutrition supervisor**: A user who have the privilege to approve food additions done by registered users.
- **IT administrator**: A user who have the privilege to monitor reviews written by registered users, and look after the system in general.

All types of users can conduct a search by name or by barcode for a food and can retrieve its information in separate page. Food information page contains: food's name, food's picture, number of calories in a portion of this food, and food's composition content table; if available. Note that food portions can be cups, plates, slices, 100 grams, etc., and food's composition content table is the nutrient data in this food like protein, fat, vitamins, etc. [19].

Guest users can register in the system to access the full features of the application. To register in the system users should provide a username, a password and a phone number. Users can optionally provide their height, waist, hips, weight, date of birth, and gender, to be used later to construct diet plan.

Registered users can write reviews for foods. They can delete their own reviews. Users also can like/dislike others reviews to express their agreement or disagreement with these reviews. Moreover, registered users can add new foods to feed the application database. In case of food addition, registered users should provide full food information same as the one presented in the food information page. The addition request should be forwarded by the system to the nutrition supervisor for approving. If approved, then, the addition request will be forwarded by the system to the IT administrator to add it to the system database. The supervisor and the administrator should be notified by an SMS if new requests are arrived and need to be processed.

Most importantly, registered uses can create, delete, and manage a diet plan. If users choose to construct a diet plan then they will be requested to input their height, weight, waist, hips, date of birth, and gender; if they have not input them in the registration. Depending on these information the system will firstly, determine the user body mass index (BMI), i.e., a measurement of the user body fat based on the weight in relation to the height. The value of the BMI is used to categorize users as underweight, normal weight, overweight, or obese [19]. The waist and hips measurements are used to calculate users abdominal fats [19]. Secondly, the system will calculate the minimum and maximum number of calories that a user can take per day. The user will be requested to choose a target number of calories in the range between the minimum and maximum. After that, the system will monitor user progress towards losing weight by calculating the number of gained calories recoded daily by the user, and compare it to the target number of calories chosen by the user. In addition, the system tracks users water drinking by requesting users to log their daily drinking amount.

The IT administrator is the person who looks after the system from the technical point of view. In addition, the administrator is responsible of monitoring users reviews and delete inappropriate ones if found.

1) Strong Aspects: A research was conducted in 2015 [20] to define the features of the ideal weight loss applications depending on the insights of Saudi women. The research concludes that the suggested themes for an ideal weight loss application include: Arabic language, culturally sensitive, motivational support and social networking, and user-friendly interface.

So'rah supports Arabic language and is designed for Saudi society. Moreover, the application provides a review section which can be used by fellow users to exchange experiences. *So'rah* also implements an easy to use interface which requires no special technical skills. According to this, *So'rah* could be considered as an ideal weight loss application for Saudi women.

Another strong point is that *So'rah* requires no registration to search its food lists. This could result in more potential users for our system; users who want to register and users who want not to register. Most of the available weight loss applications require registration before using any of their features.

Bearing in mind that an email is not essential in Saudi Arabia to get governmental services, some Saudis, in particular elders, do not have email accounts but they have phone numbers. For that, *So'rah* confirms registration via SMS mobile messages instead of emails. This could result in more potential users for our system. Most of the available weight loss applications confirm registration via emails.

In addition to these aspects, *So'rah* application presents several strong points and novelties comparing to current Arabic

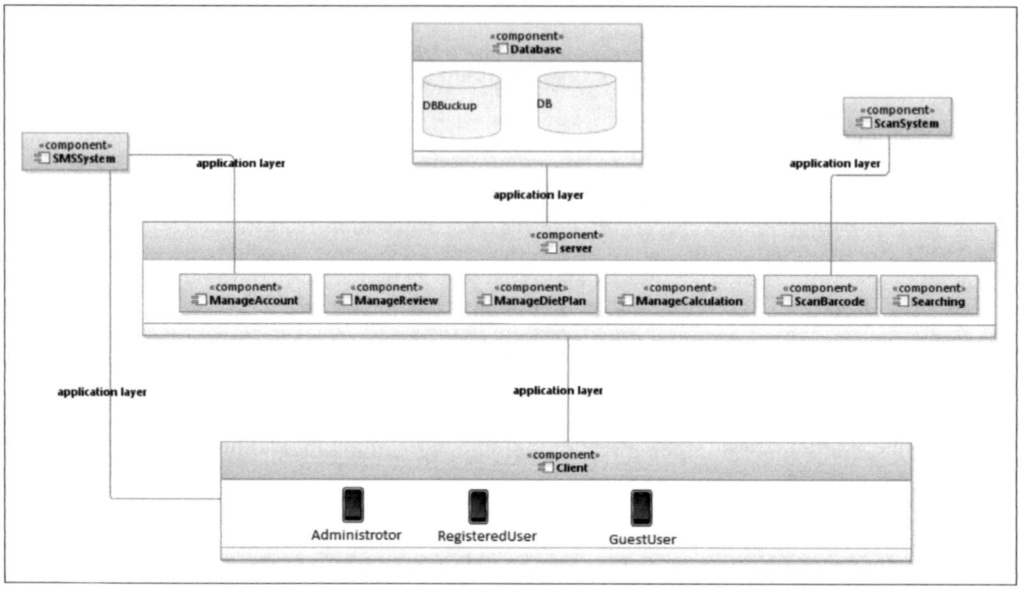

Figure 2. The System Architecture

dietary applications as been previously mentioned in different sections of this paper.

B. *The System Architectural Design and Implementation*

Considering the system programmers experiences and known that most Saudi participants [20] own Android mobile phone, the application was designed and implemented to target mobile devices running Android platforms. However the system can also be reproduced to other mobile operating systems, such as iPhone, Windows mobile, BlackBerry, etc.

The system was implemented using the Android Software Development Kit (SDK) [21]. SDK provides the needed tools and APIs to create applications for Android platform using Java programming language [22].

The Java programming language is an object oriented language, i.e., it depends on having separate entities (objects) which encapsulate the code of their functions and variables [22]. For that, an object oriented approach was followed in the design process of *So'rah*. We developed a *use case* diagram and a *class* diagram for our system with all the supporting information. Presenting the detailed design process for the application is out of the scope of this paper; interested reader can refer to [23] for information about object oriented design. The design is then mapped to Java programming language for implementation.

So'rah system architecture which identifies the system core components and their relationship is presented in Figure 2. As may be seen in the figure we choose the client server architecture because we have subsystem provides services to instances of other subsystems. The subsystem that provides services called server, the subsystem that consumed services called client.

This separation of the applications services into client/server components allows these components to be distributed in different physical machines. Additionally, this separation allows each of these components to be updated without affecting other components [23].

IV. APPLICATION DEMONSTRATION AND VALIDATION

This section presents an overview of the mobile application from the user perspective and demonstrates its use in a practical deployment.

A. *So'rah Demonstration*

As mentioned in Section III-A, we adopt a simple user interface. The interface uses buttons with considerable size to be appropriate for fingers use. The application core function is to show the number of calories in foods either by scanning the food barcode or by searching the application's database using food's name. Therefore, the main screen of the application shows these two options very simply, as shown in Figure 3. Figure 4 shows the steps to search by food's name while Figure 5 shows the steps to scan a food's barcode.

In case of searching by food's name, keywords are enough for searching, the exact name of a food is not required. If a part of the food's name is typed, then the application will give a list of suggested food items, as shown in step 3 of Figure 4.

If the intended food is allocated then a separate screen will be displayed to show the food's information. The screen has a button to display the reviews on this food, as illustrated in Figure 6.

Users are not require to register in the application to be able to search, to scan barcodes, or to view reviews. However, if users register in the application they will have extra features to boost their plan in loosing weight and to participate in food reviewing and addition.

To register in the application users can press the registration button in the main screen. Users then will be asked to provide a suggested user name, a password, and a telephone number to complete the registration, as shown in Figure 7. In addition,

Figure 3. *So'rah* Main Screen

Figure 5. Search by Food's Barcode

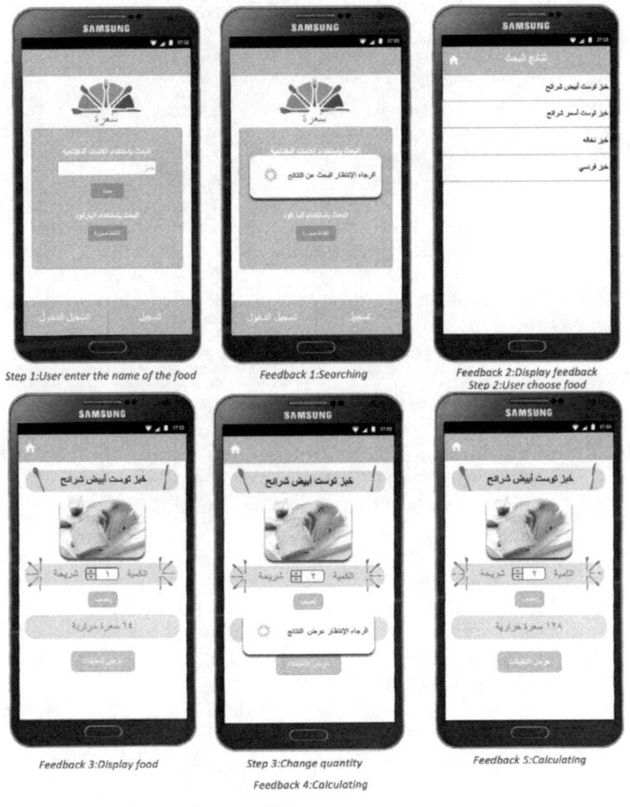

Figure 4. Searching by Food's Name

Figure 6. View Reviews

Figure 7. Registration Process

users can provide their hight, weight, waist, hips, gender, and birthday. The extra information are used to calculate the user body fat and abdominal fat. Registered users can construct a diet plan to follow up their weight by pressing *construct a diet plan* button from the user profile screen, as shown in Figure 8. If users have not provide the extra information (hight, weight, waist, hips, gender, and birthday) during registration then they will be asked to provide them. According to these information, the minimum and maximum daily calories to be consumed are calculated in addition to the body fat and abdominal fat.

Users after that will enter their consumed food for each meal in the day. The application will calculate the gained calories and compare them to the target daily calories. The system will follow the users progress in loosing weight. Users' weight will be updated according to the daily calculation and

Figure 8. Construct a Diet Plan

users will be informed if they reach the ideal weight.

Registered users can also add reviews, delete their previously written reviews, and like/dislike others reviews.

1) The Application Validation: The application functions and interfaces were preliminary tested and validated by a formal committee from the Software Engineering Department at King Saud University. In addition, an informal testing on a small group of random users was conducted to preliminary investigate the application usability.

Currently, *So'rah* team and with the collaboration of the Department of Food Science and Nutrition at King Saud University is carrying a formal testing to investigate the effectiveness of *So'rah* application in losing weight and its usability. However, the result is not available yet.

V. Conclusion and Future Work

In the paper, we present the first version of *So'rah* application. We first question the need for such an application. After that, we review the related works and highlight how our application is different. Finally, we demonstrate the application design and implantation process.

The first version of *So'rah* concerns calculating the gained calories from consumed food. Currently, we are in the design phase for the second version of *So'rah* which enhances the functionality of the first version by adding the calculation for burned calories during exercises.

Additionally, in the meanwhile we investigating the effectiveness of *So'rah* application in helping users to lose weight.

In the future, we are planing to enrich the application features by classifying foods according to their suitability for some medical diseases like kidney diseases, diabetes, high blood pressure, etc.

Acknowledgment

The authors would like to thank Dr. Ghalia H. Shamlan from the Department of Food Science and Nutrition at King Saud University for providing *So'rah* team with the references to nutrition information of Saudi foods. Moreover, we would like to thank her for dedicating some of her valuable time to answer our nutrition questions.

References

[1] E. DeNicola, O. S. Aburizaiza, A. Siddique, H. Khwaja, and D. O. Carpenter, "Obesity and public health in the Kingdom of Saudi Arabia," Reviews on Environmental Health, vol. 30, no. 3, 2015, pp. 191–205.

[2] S. Alqarni, "A review of prevalence of obesity in saudi arabia," Journal of Obesity & Eating Disorders, vol. 02, no. 2.

[3] D. Withrow and D. A. Alter, "The economic burden of obesity worldwide: a systematic review of the direct costs of obesity," Obesity Reviews, vol. 12, no. 2.

[4] World Health Organisation, "Obesity and overweight Fact Sheet N°311," 2018, accessed = 2018-2-10. [Online]. Available: http://www.who.int/mediacentre/factsheets/fs311/en/

[5] Z. A. Memish, C. El Bcheraoui, M. Tuffaha, M. Robinson, F. Daoud, S. Jaber, S. Mikhitarian, M. Al Saeedi, M. A. AlMazroa, A. H. Mokdad, and A. A. Al Rabeeah, "Obesity and associated factors - kingdom of saudi arabia, 2013," Preventing Chronic Disease, vol. 11, no. 140236.

[6] B. Spring, J. M. Duncan, E. A. Janke, A. T. Kozak, H. G. McFadden, A. DeMott, P. Alex, L. H. Epstein, J. Siddique, C. A. Pellegrini, J. Buscemi, and D. Hedeker, "Integrating technology into standard weight loss treatment: A randomized controlled trial," JAMA Internal Medicine, vol. 173, no. 2, 2013, pp. 105–111.

[7] healthline, "The 10 Best Weight Loss Apps That Help You Shed Pounds," accessed = 2018-2-10. [Online]. Available: https://www.healthline.com/nutrition/10-best-weight-loss-apps

[8] Gotta Be Mobile, "14 Best Weight Loss Apps for 2018," accessed = 2018-2-14. [Online]. Available: https://www.gottabemobile.com/best-weight-loss-apps/

[9] Red Book magazine, "The Best Weight Loss Apps You Need to Download to See Real Results in 2018," accessed = 2018-2-10. [Online]. Available: https://www.redbookmag.com/body/health-fitness/g3162/weight-loss-apps/

[10] Paste magazine, "The 10 Best Apps to Help You Eat Healthy and Lose Weight," accessed = 2018-2-14. [Online]. Available: https://www.pastemagazine.com/articles/2017/06/the-10-best-apps-to-help-you-eat-healthy-and-lose.html

[11] Lose It, accessed = 2018-2-14. [Online]. Available: https://www.loseit.com/

[12] MyFitnessPal, accessed = 2018-2-14. [Online]. Available: https://www.myfitnesspal.com/

[13] FatSecret, accessed = 2018-2-14. [Online]. Available: https://www.fatsecret.com/

[14] Weight Watchers, accessed = 2018-2-14. [Online]. Available: https://www.weightwatchers.com/us/

[15] Soraate, accessed = 2018-2-14. [Online]. Available: http://www.soraate.com/

[16] Calorie counter of my diet, accessed = 2018-2-14. [Online]. Available: https://play.google.com/store/apps/details?id=com.mdiet.tracker&hl=ar

[17] Calories guide, accessed = 2018-2-14. [Online]. Available: https://play.google.com/store/apps/details?id=com.androyal.caloriesguide.ar&hl=en

[18] My diet, accessed = 2018-2-14. [Online]. Available: http://www.m-diet.com/Home/Tracker

[19] F. S. Webb and E. N. Whitney, Nutrition concepts and controversies , 11th ed. Belmont, CA Thomson/ Wadsworth, 2008, includes bibliographical references and index. [Online]. Available: http://openlibrary.org/books/OL22568762M

[20] A. A. Alnasser, S. A. Alkhalifa, A. Sathiaseelan, and D. Marais, "What Overweight Women Want From a Weight Loss App: A Qualitative Study on Arabic Women," JMIR mHealth uHealth, vol. 3, no. 2, 2015, p. e41.

[21] N. Smyth, Android Studio Development Essentials, 1st ed. USA: CreateSpace Independent Publishing Platform, 2014.

[22] K. Arnold, J. Gosling, and D. Holmes, The Java Programming Language, 3rd ed. Boston, MA, USA: Addison-Wesley Longman Publishing Co., Inc., 2000.

[23] I. Sommerville, Software Engineering, 9th ed. USA: Addison-Wesley Publishing Company, 2010.